Radio Production

Fourth Edition

Radio Production
A manual for broadcasters

Fourth Edition

Robert McLeish

AMSTERDAM • BOSTON • HEIDELBERG • LONDON • NEW YORK • OXFORD
PARIS • SAN DIEGO • SAN FRANCISCO • SINGAPORE • SYDNEY • TOKYO

ELSEVIER

Focal Press is an imprint of Elsevier

Focal Press
An imprint of Elsevier Science
Linacre House, Jordan Hill, Oxford OX2 8DP
200 Wheeler Road, Burlington MA 01803

First published as *The Technique of Radio Production* 1978
Reprinted 1982, 1984, 1986
Second edition 1988
Reprinted 1989
Third edition 1994
Reprinted 1996, 1997, 1998
Fourth edition 1999
Reprinted 2000, (twice), 2001, 2002, 2003, 2004

British Library Cataloguing in Publication Data
McLeish, Robert
    Radio Production: a manual for broadcasters – 4th ed.
    1.   Radio – production and direction
    I.   Title
    791.4'4'0232

Library of Congress Cataloguing in Publication Data
McLeish, Robert
    Radio Production: a manual for broadcasters/Robert McLeish/4th ed
    p. cm.
    Includes bibliographical references and index
    1.   Radio broadcasting   2.   Radio plays – production and direction
    I.   McLeish, Robert   Radio Production   II.   Title
    PN1991.75.M33 1999

ISBN 0 240 51554 4

> For information on all Focal Press publications
> visit on website at www.focalpress.com

Typeset at Replika Press Pvt Ltd, Delhi 110 040, India
Printed and bound in Great Britain

Contents

Preface to the fourth edition

The world is changing faster than ever, and the world of radio with it. The boundaries of broadcasting have become blurred by the new technologies so that its delivery is no longer confined to terrestrial transmitters but includes satellites, cable, and the Internet as a means of supplying programmes on demand. The proliferation of outlets – the expansion of choice – has led to greatly increased competition, bringing with it new economic pressures and constraints. Ratings and reach become ever more important for commercial and public service stations alike as their managements increasingly seek to justify their output in purely financial terms.

Across the Middle East, Africa, and Asia, the largely unsuccessful attempts by governments to ban satellite dishes has led to the deregulation of broadcasting in country after country. If you are forced to have an 'open skies' policy it is at least better to have programmes made where officialdom can keep an eye on them – besides, the licensing of private stations is another means of raising government income.

Digital technologies continue to make equipment not only smaller and lighter but immensely more versatile so that the facilities of whole studios – sourcing, presenting, interviewing, mixing, multi-track recording and editing – can be compressed into a portable package for automated playout. Create the same techniques for pictures as for sounds, and there is no technical reason why they should not be crafted by the same people. Multi-skilling reduces staff costs and it is part of 'convergence' whereby more programmes can be produced by fewer people. And who is to say that this is not the right way to go? Broadcasting has to survive alongside far more competitors, not to say predators, than it knew in former years. So the major trends in broadcasting continue to be shaped as much by politics and economics as by its technologies, and as the clamour of communication increases so must its clarity, transparency, and appeal to personal need.

For this fourth edition my research has been more broadly international, reflecting radio practice from Haiti to Hong Kong, from the major London independents to programme suppliers in Zimbabwe, India, and Singapore. And everywhere, traditional production technique has given way to a huge variety of methods. Where once $\frac{1}{4}$-inch tape was king, we now record on minidiscs, computer hard disk, DAT cassettes, laptop floppies, analogue audio cassettes, CDs, VHS cassettes, and flash cards using recorders with no moving parts – all this in a system where the radio may be driven by wind-up clockwork! We have a powerful new ideas source and research tool in the Internet, which also illustrates an explosive divergence in the details of programme making. Because there are so many ways of achieving the end product, the technical notes here – mostly found in Chapter 2 – are reflections of a colourful mosaic of current practice. But while production method has energetically diversified, the editorial process – and that is what this book is essentially about – has remained constant in its best values of integrity, imagination, accuracy, and truth.

The elements of a good interview or an absorbing piece of drama, the points to remember in preparing a sports commentary or writing a feature, the principles of news presentation or how I should plan my programmes – these are the issues we want to address here. And going on from there to even more important questions: What is my motivation to be in radio? What is to be my relationship with the listener? Do I want to be the listener's advising friend or his counterfeit cousin, the propagandist big brother? Put like this the answer may seem obvious, but many stations with the highest motives invest in half-truths on the grounds that it is in the best interests of their audience.

Incidentally, if I too often refer to the listener or broadcaster as 'he', I apologise but there is not one instance which excludes the feminine. Both genders are intended even if not explicit in the construction.

In personal terms, I must acknowledge my debt to the late Frank Gillard for his encouragement and advice with the original draft, as well as to all those broadcasters around the world who have contributed to this new edition. Stations like BBC Radio Cambridgeshire, the ABC's smart little optout 5CK at Port Pirie, AWR in Russia, FEBA Radio in Harare, and the Journalist Training Project in Ethiopia. Thanks too for Dave Wilkinson's ideas for many of the original illustrations and to Jeff Link, a producer in BBC Training and Development who has again been invaluable in the updating process. My colleagues at ICTI, the International Communication Training Institute, and fellow trainers – Phill Butler and Walt Winters in the US, Dr Menkir Esayas in Nairobi, Frank Gray of FEBC, Dr Ross James in Australia, Dick and Flora Davies and Christopher Singh of Radio Worldwide – they speak words of remembered wisdom, as so often do the students. Dr Graham Mytton, an audience research consultant formerly with the BBC's World Service, helpfully revised the section on programme evaluation. I must also thank the copyright owners of the works attributed in these pages, in particular the radio commercials.

Revising and updating a book is far more difficult than writing one from scratch. Producers will know that the challenge of a blank sheet is more exciting and often easier to deal with than altering an existing script. I therefore willingly thank my wife for once again working with me through this process. Her clear, not to say blunt, advice has been what all producers need in their times of doubt – the voice of the listener.

Robert McLeish

Characteristics of the medium

From its first tentative experiments and the early days of wireless, radio has expanded into an almost universal medium of communication. It leaps around the world on short waves linking the continents in a fraction of a second. It jumps to high satellites to put its footprint across a quarter of the globe. It brings that world to those who cannot read and helps maintain a contact for those who cannot see.

It is used by armies in war and by amateurs for fun. It controls the air lanes and directs the taxi. It is the enabler of business and commerce, the essential for fire brigades and police, the commonplace of the mobile phone. Broadcasters pour out thousands of words every minute in an effort to inform, educate and entertain, propagandise and persuade; music fills the air. Community radio makes broadcasters out of listeners and the Citizen Band gives transmitter power to the individual.

Whatever else can be said of the medium, it is plentiful. It has lost the sense of awe which attended its early years, becoming instead a very ordinary and 'unspecial' method of communication. To use it well we may have to adapt the formal 'written' language which we learnt at school and rediscover our oral traditions. How the world might have been different had Guglielmo Marconi lived before Johann Gutenberg.

To succeed in a highly competitive marketplace where television, lifestyle magazines, newspapers, cinema, theatre, videos, websites and CDs jostle for the attention of a media-conscious public the radio producer must first understand the strengths and weaknesses of his medium.

Radio makes pictures

It is a blind medium but one which can stimulate the imagination so that as soon as a voice comes out of the loudspeaker, the listener attempts to visualise what he hears and to create in the mind's eye the owner of the voice. What pictures are created when the voice carries an emotional content – an interview with wives gathering at a pit head after news of a mining accident – the halting joy of relatives on opposite sides of the world linked by a DJ programme.

Unlike television where the pictures are limited by the size of the screen, radio's pictures are any size you care to make them. For the writer of radio drama it is easy to involve us in a battle between goblins and giants, or to have our spaceship land on a strange and distant planet. Created by appropriate sound effects and supported by the right music virtually any situation can be brought to us. As the schoolboy said when asked about television drama, 'I prefer radio, the scenery is so much better'.

But is it more accurate? Naturally, a visual medium has an advantage when demonstrating a procedure or technique, and a simple graph is worth many words of explanation. In reporting an event there is much to be said for seeing film of say a public demonstration rather than leaving it to our imagination. Both sound and vision are liable to the distortions of selectivity, and in news reporting it is up to the integrity of the individual on the spot to produce as fair, honest and factual an account as possible. In the case of radio, its great strength of appealing directly to the imagination must not become the weakness of allowing individual interpretation of a factual event, let alone the deliberate exaggeration of that event by the broadcaster. The radio writer and commentator chooses his words so that they create appropriate pictures in the listener's mind, and by so doing he makes his subject understood and its occasion memorable.

Radio speaks to millions

Radio is one of the mass media. The very term broadcasting indicates a wide scattering of the output covering every home, village, town, city, and country within the range of the transmitter. Its *potential* for communication therefore is very great but the *actual* effect may be quite small. The difference between potential and actual will depend on matters to which this book is dedicated – programme relevance, excellence and creativity, operational competence, technical reliability, and consistency of the received signal. It will also be affected by the size and strength of the competition in its many forms. Broadcasters sometimes forget that people have other things to do – life is not all about listening to radio and watching television.

Audience researchers talk about *share* and *reach*. Audience share is the amount of time spent listening to a particular station, expressed as a percentage of the total radio listening in its area. Audience reach is the number of people who *do* listen to something from the station over the period of a day or week, expressed as a percentage of the total population who *could* listen. Both figures are significant. A station in a highly competitive environment may have quite a small share of the total listening, but if it manages to build a substantial following to even one of its programmes, let alone the aggregate of several minorities, it will enjoy a large reach. The mass media should always be interested in reach.

Radio speaks to the individual

Unlike television where the viewer is observing something coming out of a box 'over there', the sights and sounds of radio are created within us, and can have greater impact and involvement. Radio on headphones happens literally inside your head. Television is in general watched by small groups of people and the reaction to a programme is often affected by the reaction between individuals. Radio is much more a personal thing, coming direct to the listener. There are obvious exceptions: in the rural areas of less developed countries a whole village will gather round the set. However, even here, the transistor revolution has made a radio an everyday personal item.

The broadcaster should not abuse this directness of the medium by regarding the microphone as an input to a public address system, but rather a means of talking directly to the individual listener. If the programme is transmitted 'live', then the broadcaster has the further advantage of an immediate link with the individual and with thousands like him. The recorded programme introduces a shift in time and, like a newspaper, is capable of being out of date. The *medium,* however, is one to one, and 'now'.

The speed of radio

Technically uncumbersome, the medium is enormously flexible and is often at its best in the totally immediate 'live' situation. No processing of film, no waiting for the presses. A report from a correspondent overseas, a listener talking on the phone, the radio car in the suburbs, a sports result from the local stadium, a concert from the capital, are all examples of the immediacy of radio. This ability to move about geographically generates its own excitement. This facility of course is long since regarded as a commonplace, both for television and radio. Pictures and sounds are bounced around the world, bringing any event anywhere to our immediate notice. Radio speeds up the dissemination of information so that everyone – the leaders and the led – knows of the same news event, the same political idea, declaration or threat. If knowledge is power, radio gives power to us all whether we exercise authority or not.

Radio has no boundaries

Books and magazines can be stopped at national frontiers but radio is no respecter of territorial limits. Its signals clear mountain barriers and cross ocean deeps. Radio can bring together those separated by geography or nationality – it can help to close other distances of culture, learning or status. The programmes of political propagandists or of Christian missionaries can be sent in one country and heard in another. Sometimes met with hostile jamming, sometimes welcomed as a life-sustaining truth, programmes have a liberty independent of lines on a map. Obeying the rules of transmitter power, sunspot activity, channel interference and receiver sensitivity, radio can bring freedom to the oppressed and enlightenment to those in darkness.

The simplicity of radio

The basic unit comprises one person with a microphone and recorder rather than a crew with camera, lights and sound recorder. This makes it easier for the non-professional to take part, thereby creating a greater possibility for public access to the medium. In any case, sound is better understood than vision, with cassette players and stereo equipment found in most schools and homes. It is also probably true that whereas with television or print any loss of technical standards becomes immediately obvious and unacceptable, with radio there is a recognisable margin between the excellent and the adequate. This is not to say that one should not continually strive for the highest possible radio standards.

For the broadcaster, radio's comparative simplicity means a flexibility in its scheduling. Items within programmes, or even whole programmes, can be dropped to be replaced at short notice by something more urgent.

Radio is cheap

Relative to the other media, both its capital cost and its running expenses are small. As broadcasters round the world have discovered, the main difficulty in setting up a station is often not financial but lies in obtaining a transmission frequency. Such frequencies are safeguarded by governments as signatories to international agreements and are not easily assigned.

Radio stations are financed in various ways including public licence, commercial advertising, government grant, private capital, public subscription, or any of these methods in combination.

Because the medium is cheap to use and can attract a substantial audience the cost per hour – or more significantly the cost per listener hour – is low. Such figures have to be provided for advertisers, sponsors, supporters and accountants. But it is also important for the producer as well as the executive manager to know what a programme costs relative to its audience. This is not to say that cost-effectiveness is the only measure of worth – it most certainly is not – but it is one of the factors which inform scheduling decisions.

The relatively low cost once again means that the medium is ideal for use by the non-professional. Because time is not so expensive or so rare, radio stations are encouraged to take a few gambles in programming.

Radio is a commodity which cannot be hoarded, neither is it so special that it cannot be used by anyone with something interesting to say. Through all sorts of methods of listener participation, the medium is capable of offering a role as a two-way communicator, particularly in the area of community broadcasting.

Radio is also cheap for the listener. The development of printed circuit boards and solid state technology allows sets to be mass produced at a cost which enables their virtual total distribution. More affordable than a set of books, good radio brings its own 'library' which is of especial value to those who cannot read – the illiterate, the blind, the person who for whatever reason is deprived of literature in his own language. The broadcaster should never forget that while he may regard his own installations (studios, transmitters, etc.) as expensive, the greater part of the total capital cost of any broadcasting system is borne directly by the public in buying receivers.

The transient nature of radio

It is a very ephemeral medium and if the listener is not in time for the news bulletin, it is gone and he has to wait for the next. Unlike the newspaper which he can put down, come back to or pass round, broadcasting imposes a strict discipline of having to be there at the right time. The radio producer must recognise that while he may store his programme in the archives, his work is only short-lived for the listener. This is not to say that it may not be memorable, but the memory is fallible and without a written record it is easy to be misquoted or taken out of context. For this reason it is often advisable for the broadcaster to have some form of audio or written log as a check on what was said, and by whom. In some cases this may be a statutory requirement of a radio station as part of its public accountability. Where this is not so, lawyers have been known to argue that it is better to have no record of what was said – for example, in a public phone-in. Practice would suggest, however, that the keeping of a recording of the transmission is a useful safeguard against allegations of malpractice, particularly from complainants who missed the broadcast and who heard about it at second-hand.

The transitory nature of the medium also means that the listener must not only hear the programme at the time of its broadcast, but must also understand it then. The impact and intelligibility of the spoken word should occur on hearing it – there is seldom a second chance. The producer must therefore strive for the utmost logic and order in the presentation of his ideas, and the use of clearly understood language.

Radio as background

Radio allows a more tenuous link with its user than that insisted upon by television or print. The medium is less demanding in that it permits us to do other things at the same time, programmes become an accompaniment to something else. We read with music on, eat to a news magazine, or hang wallpaper while listening to a play. Radio suffers from its own generosity – it is easily interruptible. Television is more complete, taking our whole attention, 'spoonfeeding' without demanding effort or response, and tending to be compulsive at a far lower level of interest than radio requires of its audience.

Because radio is so often used as background, it frequently results in a low level of commitment on the part of the listener. If the broadcaster really wants the listener to *do* something – to act – then radio should be used in conjunction with another medium. Educational broadcasting, for example, needs printed fact-sheets, booklet material, and tutor hot-lines involving schools or universities. Radio evangelism has to be linked with follow-up correspondence and involve local churches or on-the-ground missionaries. Advertising requires appropriate recall and point of sale material. While radio can claim some spectacular individual action results, in general producers have to work very hard to retain their part-share of the listener's attention.

Radio is selective

There is a different kind of responsibility on the broadcaster from that of the newspaper editor in that the radio producer selects exactly what is to be received by his consumer. In print, a large number of news stories, articles and other features are set out across several pages. Each one is headlined or identified in some way to make for easy selection. The reader scans the pages choosing to read those items which interest him – he is using his own judgement. With radio this is not possible. The selection process takes place in the studio and the listener is presented with a single thread of material, it is a linear medium. Choice for the listener exists only in the mental switching-off which occurs during an item which fails to maintain his interest, or when he tunes to another station. In this respect a channel of radio or television is rather more autocratic than a newspaper.

Radio lacks space

A newspaper may carry 30 or 40 columns of news copy – a 10-minute radio bulletin is equivalent to a mere column and a half. Again, the selection and shaping of the spoken material has to be tighter and more logical. Papers can devote large amounts of space to advertisements, particularly to the 'small ads', and personal announcements such as births, deaths and marriages. This is ideal scanning material but it is not possible to provide such detailed coverage in a radio programme.

A newspaper is able to give an important item additional impact simply by using more space. The big story is run using large headlines – the picture is blown up and splashed across the front page. The equivalent in a radio bulletin is to lead with the story and to illustrate it with a voice report or interview. There is a tendency for everything in the broadcast media to come out of the set the same size. An item may be run longer but this is not necessarily the same as 'bigger'. Coverage described as 'in depth' may only be 'at length'. There is limited scope for indicating the differing importance of an economic crisis, a religious item, a murder, the arrival of a pop group, the market prices and the weather forecast. It could be argued that the press is more likely to use this ability to emphasise certain stories to impose its own value judgements on the consumer. This naturally depends on the policy of the individual newspaper editor. The radio producer is denied the same freedom of manoeuvre and this can lead to the feeling that all subjects are treated in the same way, a criticism of bland superficiality not infrequently heard. On the other hand, this characteristic of radio perhaps restores the balance of democracy, imposing less on the listener and allowing him to make up his own mind as to what is important.

The personality of radio

A great advantage of an aural medium over print lies in the sound of the human voice – the warmth, the compassion, the anger, the pain and the laughter. A voice is capable of conveying much more than reported speech. It has inflection and accent, hesitation and pause, a variety of emphasis and speed. The information which a speaker imparts is to do with the style of presentation as much as the content of what is said. The vitality of radio depends on the diversity of voices which it uses and the extent to which it allows the colourful turn of phrase and the local idiom.

It is important that all kinds of voices are heard and not just those of professional broadcasters, power holders and articulate spokesmen. The technicalities of the medium must not deter people in all walks of life from expressing themselves with a naturalness and sincerity which reflects their true personalities. Here radio, uncluttered by the pictures which accompany the talk of television, is capable of great sensitivity, and of engendering great trust.

Radio teaches

Radio works particularly well in the world of ideas. As a medium of education it excels with concepts as well as facts. From dramatically illustrating an event in history to pursuing current political thought it has a capability with any subject that can be discussed, taking the learner at a predetermined pace through a given body of knowledge. With musical appreciation and language teaching it is totally at home. Of course it lacks television's ability to demonstrate and show, it does not have charts and graphs – as a medium it is more literate than numerate – but backed up by a teacher's notes even these limitations can be overcome and a booklet helps to give memory to the understanding. Add the correspondence element and you have the two-way questioning process which is at the heart of all personal learning.

From Australia's 'School of the Air' to the UK's 'Open University', radio effectively reaches out to meet the formal and informal learning needs of people who want to grow.

Radio has music

Here are the Beethoven symphonies, the top 40, tunes of our childhood, jazz, opera, rock and favourite shows. From the best on CD to a quite passable local church organist, radio provides the pleasantness of an unobtrusive background or the focus of our total absorption. It relaxes and stimulates inducing pleasure, nostalgia, excitement, or curiosity. The range of music is wider than the coverage of the most comprehensive record library and can therefore give the listener a chance to discover new or unfamiliar forms of music.

Radio can surprise

Unlike the CD we play or the book we pick up at home, selected to match our taste and feelings of the moment, music and speech on radio is chosen for us and may, if we let it, change our mood and take us out of ourselves. We can be presented with something new and enjoy a chance encounter with the unexpected. Radio should surprise. Broadcasters are tempted to think in terms of format radio where the content lies precisely between narrowly defined limits. This gives consistency and enables the listener to hear what he expects to hear, which is probably why he switched on in the first place. But radio can also provide the opportunity for innovation and experiment – a risk which producers must take if the medium is to surprise us in a way which is both creative and stimulating.

Radio can suffer from interference

While a newspaper or magazine is normally received in exactly the form in which it was published, radio has no such automatic guarantee. Short-wave transmission is obviously subject to deep fading and co-channel interference. Medium wave too, especially at night, may suffer from the intrusion of other stations. The quality of the sound received is likely to be very different in its dynamic or frequency range from the carefully produced balance heard in the studio. Even FM which can be temperamental is liable to a range of distortions, from the flutter caused by a passing aeroplane to ignition interference from cars and other electrical equipment.

Reception in a moving vehicle can also be difficult as the signal strength varies. Digital transmission – DAB – and direct broadcasting by satellite overcome most of these problems – at a cost – but it is as well for the producer to remember that what leaves the studio is not necessarily what is heard in the possibly noisy environment of the listener. Difficult reception conditions require compelling programmes in order to retain a faithful audience.

Given the basic characteristics of the medium, how is radio to be used? What are its possibilities? Details vary across the world but broadly it functions in two main ways – it serves the individual, and operates on behalf of society as a whole.

Radio for the individual

- It diverts people from their problems and anxieties, providing relaxation and entertainment. It reduces feelings of loneliness creating a sense of companionship.
- It is the supplier or confirmer of local, national, or international news and establishes for us a current agenda of issues and events. It satisfies our personal curiosity about what is going on.
- It offers hope and inspiration to those who feel isolated or oppressed within a restrictive environment, or where their own media are manipulated or suppressed.
- It helps to solve problems by acting as a source of information and advice. It can do this either through direct personal access to the programme or in a general way by indicating sources of further help.
- It enlarges personal 'experience', stimulating interest in previously unknown topics, events, or people. It promotes creativity and can point towards new personal activity. It meets individual needs for formal and informal education.
- It contributes to self-knowledge and awareness, offering security and support. It enables us to see ourselves in relation to others, and links individuals with leaders and 'experts'.
- It guides social behaviour, setting standards and offering role models with which to identify.
- It aids personal contacts by providing topics of conversation through shared experience, 'did you hear the programme last night?'
- It enables individuals to exercise choice, to make decisions and act as citizens, especially in a democracy through the unbiased dissemination of news and information.

Radio for society

- It acts as a multiplier of change, speeding up the process of informing a population, and heightening an awareness of key issues.
- It provides information about jobs, goods and services and so helps to shape markets by providing incentives for earning and spending.
- It acts as a watch-dog on power holders, providing contact between them and the public.
- It helps to develop agreed objectives and political choice, it enables social and political debate, exposing issues and options for action.
- It acts as a catalyst and focus for celebration – or mourning – enabling individuals to act together, forming a common consciousness.
- It contributes to the artistic and intellectual culture providing opportunities for new and established performers of all kinds.
- It disseminates ideas. These may be radical, leading to new beliefs and values, so promoting diversity and change – or they may reinforce traditional values so helping to maintain social order through the status quo.
- It enables individuals and groups to speak to each other, developing an awareness of a common membership of society.
- It mobilises public and private resources for personal or community ends, particularly in an emergency.

Some of these functions are in mutual conflict, some are applicable more to a local rather than a national community, and some apply fully only in conditions of crisis. However, the programme producer should be clear about what it is he or she is trying to achieve. Lack of clarity about a programme's purpose leads to a fuzzy, ineffective end product – it also leads to arguments in the studio over what should and should not be included. We shall return to the point, but it is not enough for the producer to want to make an excellent programme – he might as well want merely to modulate the transmitter. The question is why? What is to be its effect – on the listener, that is? Before looking at some possible personal motivations for making programmes at all, we should examine the meaning of the much-used phrase – broadcasting as a public service.

The public servant

Public service broadcasting is sometimes regarded as an alternative to commercial radio. The terms are not mutually exclusive, however; it is possible to run commercial radio as a public service, especially in near-monopoly conditions or where there is little competition for the available advertising. It is a matter of where the radio managers see their first loyalty. To put this in perspective here are ten attributes of service, using the analogy of the perfect domestic servant. He or she:

- is loyal to the employer and does not try to serve other interests or use his position of privilege for his own ends. Servants are clear about their purpose and demonstrate this in everything they do.
- understands the culture, nuances and foibles of the family he serves, and in accepting them is himself fully accepted. The servant is not critical or judgemental of those he or she serves but may challenge them, or even restrain them. But he is not a policeman, he has no prescriptive mandate.
- is available when needed, and for whoever in the family needs help, the young and the old as well as the head of the house. It may be tempting to pay court to the power holders, to keep in with the paymaster – but it could be the children, the sick, or disadvantaged who need most attention.
- is actually useful, meeting stated requirements and anticipating needs and difficulties – ready to offer original solutions to problems, as or even before they arrive. The servant looks ahead, is creative.
- is well informed, offers good advice and is able to relate unpalatable truth, risking unpopularity in the process. The servant therefore has courage as well as a well-stocked mind.
- is hard working, technically expert and efficient. Does not waste resources but is honest and if called to account for his behaviour, does so willingly. There is nothing underhand – integrity is a key attribute.
- is witty and companionable, courteous and punctual. The servant inhabits our house, comes with us in the car, and is someone with whom we can have a relationship that is enjoyable as well as professional.
- The servant, especially in a large family, has to realise that he can't do everything, that it is impossible to be in two places at once – there are priorities. The family must understand this too.
- works smoothly alongside others and does not crave recognition or status. When the number to be served is very large, other servants may need to be brought in. We may want to contract out some of the work to specialists – a governess, cook, chauffeur, gardener.
- is affordable. My servant must not bankrupt me. However much I value this service, long-term cost-effectiveness is crucial.

Each of these characteristics of service has its equivalent in broadcasting, but there is an immediate dilemma – such perfection may be too expensive. As in everything else we get the level of service we pay for and it may not be possible to afford a 24 hours a day, seven days a week output catering for all needs. Live concerts, stereo drama and world news are expensive commodities and radio managers need to make judgements about what can be provided at an acceptable price. However, the hallmarks of public service broadcasting can be drawn from our ideas of what a servant is and does. Such a station is certainly not arrogant, setting itself up as a power in its own right. It is responsive to need, making itself available for everyone – not simply the rich and powerful; indeed its universality makes a point of including the disadvantaged. It is wide ranging in its appeal, competent and reliable, entertaining and informative. Its programmes for minorities are not to be hidden away in the small hours but are part of the diversity available at prime time. It is popular in that over a period of time it reaches a significant proportion of the population. It does not 'import' its programmes but is culturally in tune with its audience, producing most of the output itself. It provides useful and necessary things – things of the quality asked for, but also unexpected pleasures. Above all, it is editorially free from interference by political, commercial or other interests, serving only one master to whom it remains essentially accountable – its public.

The concept works well for a service that is adequately paid for by its listeners – by public licence or by subscription. But if this is not the case, can a public service station do deals with third parties to raise additional revenue? Can a government, commercial or religious station be run as a genuine public service? Yes it can, but the difficulties are obvious.

First, a publicly funded service which makes arrangements with commercial interests is putting its first loyalty, its editorial integrity, at risk. Any producer making a programme as a co-production, or acting under special rules, must say so – not as part of some pay-off, but to meet the requirements of his public service accountability.

Second, there is a strong tendency for 'piper payers' to want to call the tune. A government does not like to hear criticism of its policy on a station which it regards as its own. Authority in general does not wish to be challenged – as from time to time political interviewers must. Ministries and departments are highly sensitive to items which 'in the public interest' they would rather not have broadcast. Officials will avoid or delay 'bad news', however true. Similarly, a commercial station often needs to maximise its audience in order to justify the rates, so pushing sectional interests to one side to satisfy the advertisers' desire for mass popularity. A Christian fund-raising constituency may press for the gospel *it* wants to hear, forgetting the need to serve people in a multiplicity of ways. The need to survive in a harsh political climate, or in a fiercely competitive one, exacerbates these pressures. The fact is that a station dedicated to public service but controlled or funded by a third party having its own interests to consider is almost certain to weaken in its commitment.

Types of radio station

Radio is categorised not so much by what it does as by how it is financed. Each method of funding has a direct result on the programming which a station can afford or is prepared to offer, which again is affected by the degree of competition which it faces.

The main types of organisational funding are as follows:

- Public service, funded by a licence fee and run by a national corporation.

- Commercial station financed by national and local spot advertising or sponsorship, and run as a public company.
- Government station paid for from taxation and run as a government department.
- Government-owned station, funded largely by commercial advertising, operating under a government appointed board.
- Public service, funded by government funds or grant-in-aid, run by a publicly accountable board, independent of government.
- Public service, subscription station, does not take advertising and is funded by individual subscribers and donors.
- Private ownership, funded by personal income of all kinds, e.g. commercial advertising, subscriptions, donations.
- Institutional ownership, e.g. university campus, hospital or factory radio, run and paid for by the organisation for the benefit of its students, patients, employees, etc.
- Radio organisation run for specific religious or charitable purposes – sells air time and raises income through supporter contributions.
- Community ownership financed by local advertising and sponsors.
- Restricted Service Licence – RSL stations, on low power and having a limited lifespan to meet a particular need, e.g. one-month licence to cover a city festival.

The ability to charge premium rates for advertising depends either on having a large audience in response to popular appeal programming or an audience with a high purchasing power for the individual advertiser. Competing for income from a limited supply is quite different from the competition which a station has to win an audience.

Around the world, stations combine different forms of funding and supplement their income by all possible means from profit-making events such as concerts, publishing, and the sale of programmes – both to the public on CD and cassettes as well as to other broadcasters – to the sale of T-shirts, community coffee mornings, and car boot sales.

'Outside' pressures

No radio station – and therefore no producer – exists in a vacuum. It has a context of connections, each one useful and necessary but also representing a source of potential pressure which can inhibit a single-minded commitment to the ideal of public service. Figure 1.1 illustrates some of these, each linked by two directional arrows. It is an important exercise for broadcasters to know in their own situation what these arrows represent. What are the expectations and transactions in each direction? To what extent are they fulfilled? For the individual producer the compromises and strictures of his own management are difficult enough; additional obligations to interests outside broadcasting can be extremely tiresome. Nevertheless, as a

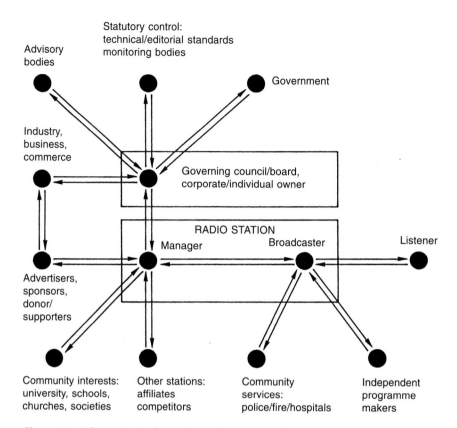

Advisory
bodies

Statutory control:
technical/editorial standards
monitoring bodies

Government

Industry,
business,
commerce

Governing council/board,
corporate/individual owner

RADIO STATION

Manager   Broadcaster   Listener

Advertisers,
sponsors,
donor/
supporters

Community interests: Other stations: Community   Independent
university, schools,   affiliates   services: programme
churches, societies   competitors   police/fire/hospitals   makers

Figure 1.1   The context of a radio station – typical links and pressure points

background to programme making it is necessary to be aware of everyone
else who has a stake in the process.

For example, the working producer should know at least two members
of the organisation's governing board or council. Not only does this help
the programme makers to appreciate the context within which the station
is operating, but also acquaints board members with the difficulties and
encouragements experienced by those who actually create the product.
Such introductions are best made through the manager.

Personal motivations

So what is our purpose in being in radio? Is it because it has some
appearance of power – able to sway public opinion and make people do
things? If so, it has to be said that this is very rare and most unlikely to
be achieved by this medium alone. Is it to be the protagonist mouthpiece
of someone else – or are there reasons which meet my own needs?

It is as well for the producer to understand what some of these personal
motivations are:

- to inform people – the role of the media journalist
- to educate – enabling people to acquire knowledge or skill
- to entertain – making people laugh, relax or pass the time agreeably
- to reassure – providing supportive companionship
- to shock – the sensation station
- to make money – a means of earning a living
- to enjoy oneself – a means of artistic expression
- to create change – crusading for a new society
- to preserve the status quo – returning to established values
- to convert to one's own belief – proselytising a faith
- to present options – allowing the listener to exercise choice.

Each programme maker must decide for himself why he is in broad-casting. It may be in order to earn a living, or because he or she has something to say. It may be out of a genuine desire to serve one's fellow men and women – to provide options for action by opening up possibilities, making them better informed.

One's success criteria are perhaps best summed up as a response to the question 'What do you want to say, to whom, and with what effect?' The 'to whom?' is important for in the end, radio is about a relationship. Much more than on television, the presenter, DJ, or newsreader establishes a sense of rapport with the listener.

The successful station is more than the sum of its programmes, it understands and values the nature of this relationship – and the role which it has for its community as both leader and servant.

2

The radio studio

Books on the technical and operational aspects of broadcasting are listed under Further Reading and the subject is not dealt with in detail here. However, to omit it altogether might appear to indicate a separation of programme production from its functional base. This can never be so. The listener is after all dependent on sound alone and must be able to hear clearly and accurately. Any sounds which are distorted, confused or poorly assembled are tiring to the listener and will not retain his interest.

The quality of the end product depends in part directly on the engineering and operational standards. It scarcely matters how good the ideas are, how brilliant the production, how polished the presentation, because all will founder on poor operational technique. Whether the broadcaster is using other operational staff, or is in the self-operational mode, a basic familiarity with the studio equipment is essential – it must be second nature.

Taking the analogy of driving a car, the good driver is not preoccupied with how to change gear, or with which foot pedal does what – he is more concerned with road sense, i.e. the position of the car on the road in relation to other vehicles. So it is with the broadcaster. His use of the tools of the trade – microphones, computers, recorders, the studio mixer, CD decks – these must all be at the command of the producer who can then concentrate on what broadcasting is really about, the communication of ideas in music and speech.

Good technique comes first, and having been mastered, should not be allowed to obtrude into the programme. The technicalities of broadcasting – editing, fading, control of levels, sound quality and so on – should be so good that they do not show. By being invisible they allow the programme content to come through.

In common with other performing arts such as film, theatre and television, the hallmark of the good communicator is that his means are not always apparent. Basic craft skills are seldom discernible except to the professional who recognises in their unobtrusiveness a mastery of the medium.

There are programme producers who declare themselves uninterested in technical things, they will leave the mixer or computer operation to others so that they can concentrate on 'higher' editorial matters.

Unfortunately if you do not know what is technically possible, then you cannot fully realise the potential of the medium. Without knowing the limitations either there can be no attempt to overcome them. You simply suffer the frustrations of always having to depend on someone else who does understand. So here, we will explore the studio and discuss some of the equipment.

Studio layout

Studios for transmission or rehearsal/recording may consist simply of a single room containing all the equipment, including one or more microphones. This arrangement is designed for use by one person and is called a self-operation or self-op studio.

Where two or more rooms are used together, the room with the mixer and other equipment is often referred to as the control room or cubicle, while the actual studio – containing mostly microphones – is used for interviewees, actors, musicians, etc. If the control cubicle also has a mic it may still be capable of self-operation. In any area, when the mic is faded up the loudspeaker is cut and monitoring must be done on headphones.

The studio desk, mixer, control panel, or board

Most studios will include some kind of audio mixer. This is essentially a device for mixing together the various programme sources, controlling their level or volume, and sending the combined output to the required destination – generally either the transmitter or a recording machine. It contains three types of circuit function.

1   *Programme circuits:* a series of audio channels, their individual volume levels controlled by separate slider faders. A second or auxiliary output – generally controlled by a small rotary fader on each channel – can provide a different mix of programme material typically used for public address, echo, foldback into the studio, a clean feed sent to a distant contributor, etc.
2   *Monitoring circuits:* a visual indication – either by a programme meter or a vertical column of lights, and an aural indication – loudspeaker or headphones, to enable the operator to hear and measure the individual sources as well as the final mixed output.
3   *Control circuits:* the means of communicating with other studios or outside broadcasts by means of 'talkback' or telephone.

In learning to operate a mixer there is little substitute for first understanding the principles of the individual equipment, then practising until its operation becomes second nature. The following are some operational points for the beginner.

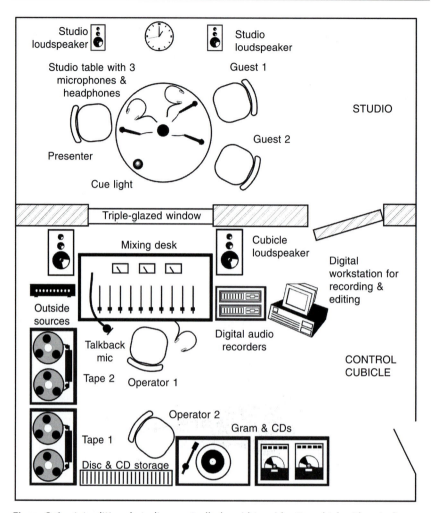

Figure 2.1 A traditional studio, controlled or 'driven' by its cubicle. The studio area contains tables and chairs with two or more microphones for interviewing, narration, or simple drama. Monitoring is by loudspeaker, or when the mic channels are open, by headphones to each person as required. The monitoring normally carries the output of the mixing desk, but it may be arranged to take any programme feed, e.g. an incoming phone call. The headphones also carry talkback from the cubicle. The cubicle contains the mixing desk with as many channels as programme complexity demands, in this case 10. They can be laid out as the operator requires, e.g. tape 1, tape 2, outside sources, mics 1, 2, 3, DAT 1, DAT 2, CD 1, CD 2. The computer may also play out through a channel. Minidisc may take the place of tape or DAT. Sources such as CDs and gram are only connected when required. The outside sources may include the newsroom, telephone, or other remote sources, studios, or networks. A switching matrix or jackfield makes each source assignable to any channel fader. A second operator is used to check and set up the tapes, minidiscs, DATs, or CDs ready to play, either on cue or by remote start activated by the fader opening. This studio is designed for the production of high-speed or complex programmes such as a lengthy news show with many speakers, guests, and recorded inserts

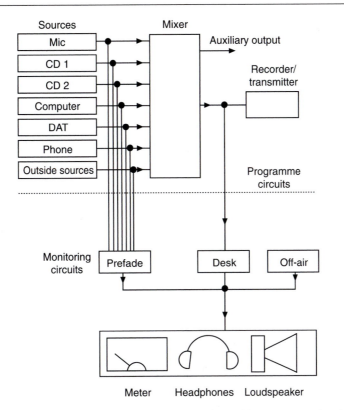

Figure 2.2 Studio mixer: typical programme and monitoring circuits showing the main programme and auxiliary outputs, and the principle of prefade, desk output, and off-air monitoring

The operator must be comfortable. The correct chair height and easy access to all necessary equipment is important for fluid operation. This mostly calls for a swivel chair on castors.

The first function to be considered is the *monitoring* of the programme. Nothing on a mixer, which might possibly be on air, should be touched until the question has been answered – What am I listening to? The loudspeaker is normally required to carry the direct output of the desk, as, for example, in the case of a rehearsal or recording. In transmission conditions it will normally take its programme off-air, although it may not be feasible to listen via a receiver when transmitting on short wave. As far as possible, the programme should be monitored as it will be heard by the listener, i.e. after the transmitter, not simply as it leaves the studio.

The volume of the monitoring loudspeaker should be adjusted to a comfortable level and then left alone. It is impossible to make subjective assessments of relative loudness within a programme if the volume of the loudspeaker is constantly being changed. If the loudspeaker has to be turned down, for example for a phone call, it should be done with a single key operation so that the original volume is easily restored afterwards.

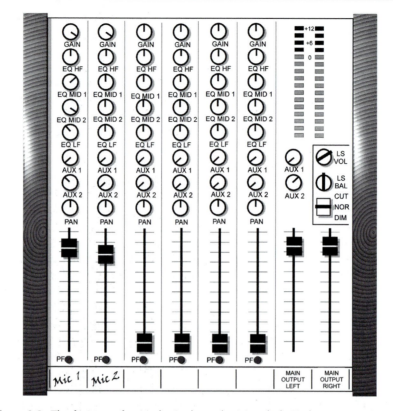

Figure 2.3 The features of a simple six-channel mixing desk. Each programme source is plugged into the mixer so that it has its own channel fader and associated controls. Each plug-in socket (not shown) has a mic/line sensitivity switch to compensate for low level inputs, e.g. from a microphone, or a high-level input, e.g. from a CD player. Practice varies, some mixers fade up by moving the fader towards the operator, but most use the near position as closed. Moving a fader off its backstop may switch on the red 'on-air' light, cut the loudspeaker, or operate 'remote start' equipment. Each channel also has:

- a prefade button, to hear and measure an individual source without fading it up
- a pan pot, for placing a source left or right in a stereo picture
- two auxiliary outputs, independent of the main programme mix
- equalisation, to cut or boost the low, middle, or high frequencies. In the mid-range one control may select the frequency while the other is used to adjust the amount of cut or boost
- a gain control, to adjust the amplifier in the channel.

A designation strip enables the operator to label each channel. The main left and right outputs have faders to adjust the overall level of the mix, as do the auxiliary outputs. The monitoring loudspeakers have their own volume control, left and right balance, dim and cut key. The output level is indicated by vertical rows of lights which change colour to red at the onset of overload distortion. In this example, mic 1 has some mid-boost 'presence' and bass roll-off. It also appears on Aux output 2. Both mics are placed centrally for a mono output but could be panned half left and half right for a stereo effect.

In *mixing* sources together – mics, CD players, computer insert, etc. – the general rule is to bring the new fader in before taking the old one out. This avoids the loss of atmosphere which occurs when all the faders are closed. A slow mix from one sound source to another is the 'crossfade'.

In assessing the relative *sound levels* of one programme source against another, either in a mix or in juxtaposition, the most important devices are the operator's own ears. The question of how loud speech should be against music depends on a variety of factors, including the nature of the programme and the probable listening conditions of the audience, as well as the type of music and the voice characteristics of the speech. There will certainly be a maximum level which can be sent to the line feeding the transmitter, and this represents the upper limit against which everything else is judged. Obviously for the orchestral concert music needs to be louder than speech. However, the reverse is the case where the speech is of predominant importance or where the music is already dynamically compressed, as it is with rock or pop. This 'speech louder than music' characteristic is general for most sequence programmes or when the music is designed for background listening. It is particularly important when the listening conditions are likely to be noisy, for example at busy domestic times or in the car.

However, it is also true, and it is especially prevalent in a situation of fiercely competing transmitters, that maximum signal penetration is obtained by sacrificing dynamic subtlety. The sound levels of all sources are kept as high as possible and the transmitter is given a large dose of compression. It is as well for the producer to know about this otherwise he may spend a good deal of time obtaining a certain kind of effect or perfecting his fades, only to have them overruled by an uncomprehending transmitter!

Probably the most important aspect of mixer operation is *self-organisation*. It is essential to have a system for handling the physical items: that is, the running order, scripts, cue sheets, CDs, and recorded items, etc. The material – and its paperwork – which has been used should be put out of the way, and new material brought forward as it is needed. The good operator is always one step ahead. He knows what he is going to do next, and having done it, sets up the next step.

Computers

Capable of recording, storing, manipulating, and replaying audio material in digital form, computers offer very high quality sound, as well as ease of editing, and immediate access to any part of the programme held. The computer in this context is used in two different ways:

1   as a 'stand-alone' terminal capable of editing, storing, or broadcasting material, but which is not connected to any other computer.

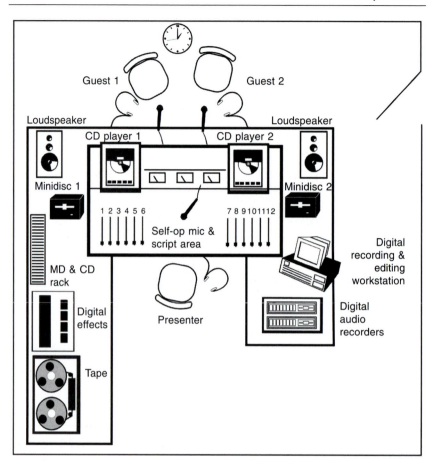

Figure 2.4   A self-op studio with 12-channel mixer and a variety of programme sources. The computer may be part of an integrated system capable of delivering audio material from a central store, or of the 'stand-alone' type able to record, or play out from its hard disk. The Effects Unit can be switched to any source, and any source connected to any fader. A typical layout would be:

1   Tape
2   MD 1
3   MD 2
4   CD 1
5   CD 2
6   Self-op mic
7   Guest mic 1
8   Guest mic 2
9   DAT 1
10   DAT 2
11   Computer o/p 1
12   Computer o/p 2

Monitoring is by stereo and prefade meters, loudspeakers and headphones

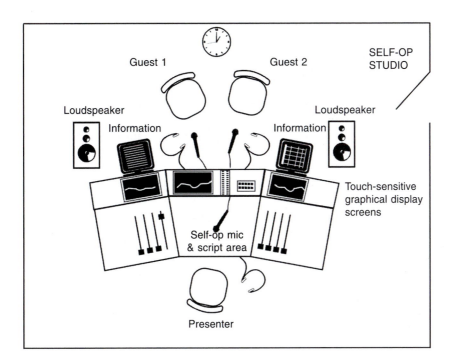

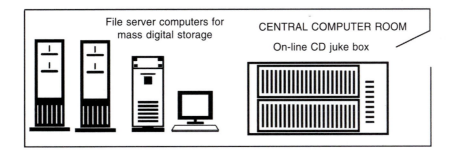

Figure 2.5 A digital self-op studio. Simplicity of operation is achieved by the removal of programme sources to a central computer room. The touch-sensitive screen icons provide variable EQ for each channel, and connect whatever is needed to each fader – the news room, pre-recorded material, CDs, outside sources, echo, sustaining network, etc. The computer can be programmed so that inputting a personal password instantly arranges the desk layout to the individual preference of the presenter. The information screens provide the programme running order, news bulletin or update, weather, details of phone-in callers, station schedule, top 40, etc. When not on-air, this arrangement is also used as a digital editing suite for making packages and recording voice-overs, etc.

2   as part of an integrated network system where all the material is
     stored centrally and may be accessed or manipulated by any one of
     a number of terminals. Individual programmes or items may be
     protected by allowing access only through a password.

The great advantage of an integrated system is that while the editing of
an item – a news report – can be done on a terminal in the newsroom, it
is then immediately available to the on-air broadcaster in the studio. The
studio terminal can have access to an immense range of material – it
simply depends on the producer understanding the system well enough to
know what is available.

In the transition to a fully computerised studio, station or network, it
is of the utmost importance that people, especially older people used to
more traditional methods of programme making, are given sufficient
time for training. The cost of this is an essential component of the total
capital investment. It takes several weeks for staff and freelances to
become familiar and comfortable with equipment of this nature. This is
partly to do with the requirement for a quite different psychological
approach. Understanding is no longer based on knowing or at least seeing
how it works – as was the case with a turntable or tape recorder. Knowing
'how' is not necessary, but knowing what it does and recognising its
potential is the new learning. Not fully understanding a new process
produces anxieties for many people, which is why no system should be
handed over to the operational users until it is thoroughly tested and
proved by the installation engineers. Once operational, the software must
be totally reliable. Even so, experience has shown that it is wise to have
a CD player or pre-recorded music on DAT easily and quickly available
on one of the mixer channels.

Digital audio workstation

This facility is likely to be in a small room separate from the studio. It
could be a 'stand-alone' terminal complete in itself, or be fully integrated
with a network computer system. The example here is a typical digital
audio workstation (Figure 2.6). It comprises an audio mixer one source
of which is the microphone, a computer and keyboard, and a unit for
picking up incoming digital ISDN lines. Alternatively, the mixer could
be supplied with feeds from the local newsroom, a remote studio, outside
broadcast, or incoming telephone call. This arrangement can therefore be
used for local voice recording, interviewing someone at a remote source,
editing material, mixing items to create a finished 'package' or correcting
levels of an already recorded mix. It is an ideal arrangement for making
short self-op news packages, especially when a digital playback machine
is included for downloading material recorded on location. (See also

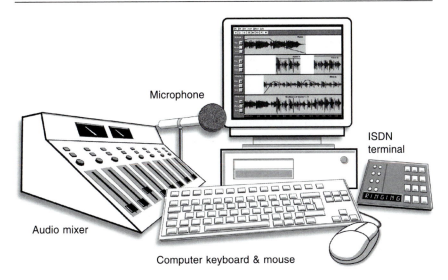

Figure 2.6 A typical digital audio workstation

Figure 7.1.) In this example we are making a programme opening, mixing together three separate sources (Figure 2.7).

Given the appropriate software, different studio items are recorded on the computer to appear as separate tracks. Here is the script we are working on.

Music	Locally recorded guitar rhythm
	(fade after 4 seconds and mix with Fx)

Fx	Surf waves on seashore
	(fade up, peak, and hold under)

Narrator   The Comores Islands. On the map they look like tiny specks in the Indian Ocean
(Slowly fade down music)
But walking along the white sandy beaches you're overwhelmed by the mass of lush green vegetation climbing up the huge mountain peaks. But they're not just mountains – they're volcanoes. Ask any passing fisherman
(music out)

FX   (Briefly up, down, hold under)

Fisherman   Oh yes – the last real eruption we had was in 1977 – Lava poured down into the sea . . .

Here, the music is put on Track 1, the speech on Track 2, and the effects on Track 3. The resulting mix appears on Track 4. At the bottom is a timescale marked off in seconds from the left.

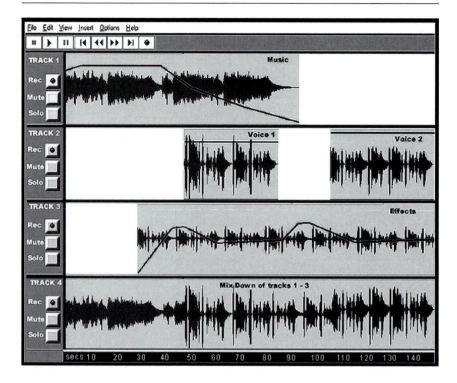

Figure 2.7 Digital editing close up. Mixing together three different sources to make a programme opening

Each of the tracks can be played independently and its level adjusted by clicking the mouse on to the volume line, or envelope, and moving it up or down. This is the solid line on each track indicating its relative level at any one time.

The music begins on its own and after 3 seconds the seawash is faded up to mix with it. After 5 seconds this mix is faded down and held under the narrator's voice. The music continues to be faded and is completely out by the time the seawash is briefly peaked before the fisherman – voice 2 – comes in. Note that the fisherman was recorded at a slightly lower level than the narrator's voice so that track 2 is brought up to compensate.

The advantage of this facility is that it allows the mixing to be repeated with 100 per cent accuracy while perhaps changing one of the levels or cues to get the desired result. It provides for split-second timing, it leaves all the original recordings intact, it doesn't tie up a whole studio, and takes only one person to achieve the final programme. If that person is the narrator who is also the producer, the production method is not only very efficient but can also create a high level of personal job satisfaction.

Tape reproduction

Magnetic tape may be used in the studio in various formats – digital audio tape (DAT), audio cassettes, $\frac{1}{4}$-inch reel to reel, and the closed cartridge. As with CDs and albums, tapes intended for broadcast should be checked before transmission to ensure that the correct tape is played. This requires the words or music at the start of the tape to be matched against the information provided on the cue sheet or running order. Since recorded levels can vary considerably, this pre-transmission check is also used to adjust the replay volume to the existing programme level.

- *Digital audio tape* – DAT – has revolutionised sound recording and editing. It is used both for the mastering of music recordings and with small portable machines (see also Minidisc) for the collection of material on location. Employing the same technology as that of video recording – a slow-speed tape scanned by a revolving head – the DAT cassette is smaller than the conventional audio cassette and, as with the video cassette, physical access to the tape is denied. The signal is recorded in digital form in which the original electrical variations are represented by a series of pulses or 'bits' of information. Audio in the form of such bits is easily manipulated by computer in the same way as typed text is re-ordered by a word processor. The sound quality remains exactly as the original and does not degrade as successive digital copies are made. Two record/replay DAT machines are often found in both transmission and recording studios – mostly for the recording or playing back of whole programmes rather than short items or programme inserts.
- *The audio cassette* uses an enclosed reel-to-reel tape 3-mm ($\frac{1}{8}$-inch) wide running at 4.7 cm ($1\frac{7}{8}$ inches) per second. Cassettes, originally developed for the domestic market, have many applications in broadcasting. The professional 'walkman' cassette recorder is easily carried and unobtrusive in operation, and while equalled in this respect by the minidisc recorder, it is the key to its popularity. However, it is wise to remember that because the mechanical tolerances on such small, relatively low-speed devices are fairly critical, there is no absolute guarantee that a cassette recorded on one machine will replay perfectly on another.
- *Reel-to-reel tape* is normally 6 mm ($\frac{1}{4}$ inch) wide and is mostly used on spools varying in diameter from 12.5 (5 inches) to 25 cm (10 inches). Playing speeds of professional machines are either 19 cm ($7\frac{1}{2}$ inches) or 38 cm (15 inches) per second. The higher the speed, the better the quality – the frequency response – and stereo recordings are typically made at twice the tape speed of mono material. Reel-to-reel tape is also used in widths of 1 and 2 inches for multi-track recording. Reel-to-reel tape is now largely superseded by digital recording, but huge stocks of library material exist on $\frac{1}{4}$-inch tape and it will therefore remain a feature of radio studios particularly as back-up and for playout.

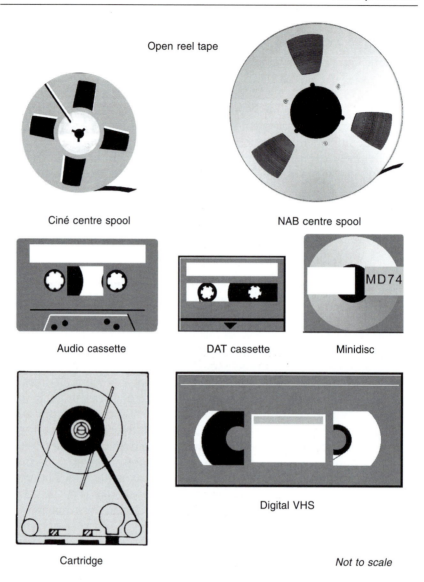

Open reel tape

Ciné centre spool   NAB centre spool

Audio cassette   DAT cassette   Minidisc

Digital VHS

Cartridge   *Not to scale*

Figure 2.8   Recording formats. Reel-to-reel tape comes in a variety of lengths and widths. Pictured here is the ciné centre spool of 12.5 cm (5 in.) and 17.5 cm (7 in.) diameter, and the 25 cm (10 in.) aluminium spool with the NAB centre. Audio cassettes are designated according to their total playing time – a C60 cassette has a 30 min duration on each side. The DAT cassette is closed like a video cassette, and has 2 hours' playing time. The minidisc has a 64 mm diameter inside a plastic case providing 140 min mono or 70 min stereo recording time. Cartridges are made of variable length from 30 seconds upwards – the specially lubricated tape is drawn from the inside of the reel and returned to the outside. The digital VHS cassette is of variable duration depending on the system used and together with the recordable CD is favoured for the storage of programme archives

Table 2.1 Running times of recording tape in hours and minutes

Tape length	Speed, cm/second (in./second)				
	38 (15)	19 (7 $\frac{1}{2}$)	9.5 (3 $\frac{3}{4}$)	4.7 (1 $\frac{7}{8}$)	2.3 ($\frac{15}{16}$)
180 metres (600 ft)	8 min	16 min	32 min	1 h 4 min	2 h 8 min
365 metres (1200 ft)	16 min	32 min	1 h 4 min	2 h 8 min	4 h 16 min
730 metres (2400 ft)	32 min	1 h 4 min	2 h 8 min	4 h 16 min	8 h 32 min

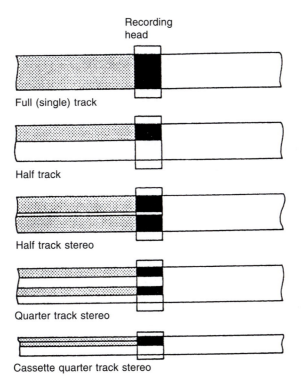

Figure 2.9 Tape track systems. Monophonically recorded cassette tape uses all the top half of the tape to ensure compatibility of playback with the stereo system. A recording system using only part of the tape enables a second signal to be recorded by 'turning the tape over', but results in a lower signal to noise ratio, i.e. greater tape hiss

- *The tape cartridge* is an enclosed single-spool device containing a continuous loop of $\frac{1}{4}$-inch tape, and is used for handling relatively short items that need to be replayed several times – signature tunes, jingles, promos, station idents, commercials, etc. This function is now more effectively done by the minidisc, or recordable CD – or simply stored on a computer storage system.

Editing principles

The purpose of editing can be summarised as:

1   To rearrange recorded material into a more logical sequence.
2   To remove the uninteresting, repetitive, or technically unacceptable.
3   To compress the material in time.
4   For creative effect to produce new juxtapositions of speech, music, sound and silence.

Editing must not be used to alter the sense of what has been said or to place the material within an unintended context.

There are always two considerations when editing, namely the editorial and the technical. In the editorial sense it is important to leave intact, for example, the view of an interviewee, and the reasons given for its support. It would be wrong to include a key statement but to omit an essential qualification through lack of time. On the other hand, facts can often be edited out and included more economically in the introductory cue material. It is often possible to remove some or all of the interviewer's questions, letting the interviewee continue. If the interviewee has a stammer, or pauses for long periods, editing can obviously remove these gaps. However, it would be unwise to remove them completely, as this may alter the nature of the individual voice. It would be positively misleading to edit pauses out of an interview where they indicate thought or hesitation.

The most frequent fault in editing is the removal of the tiny paragraph pauses which occur naturally in speech. There is little point in increasing the pace while destroying half the meaning – silence is not necessarily a negative quantity.

Editing practice

Each recording method has its own editing procedure:

* *Computer editing*   Once audio material has been downloaded on to a hard disk, it can be manipulated, cut, rearranged, or treated in a variety of ways depending on the software used. A general point of technique is almost always to cut on the beginning of a word, rather than at the end of the preceding word. This makes for a much more definite edit point.
  An advantage of computer editing over some other methods is that it leaves the original recording intact – non-destructive editing. It is therefore possible to do the same edit several times, or to try alternatives, to get it absolutely right. This is a valuable facility for editing both speech and music.

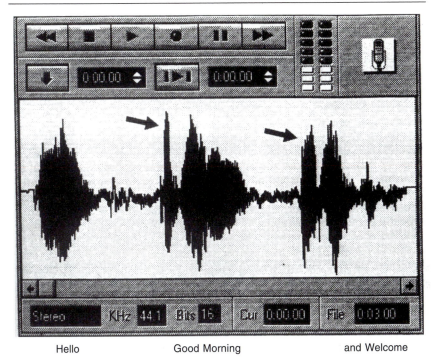

Hello   Good Morning   and Welcome

Figure 2.10 Using the mouse, the arrows have been set at the edit points in order to remove the 'Good Morning' for this programme's afternoon repeat

- *DAT* A full editing suite is expensive, requiring at least two DAT machines and an edit controller. This controller uses time-code information on the original recordings as a 'label' for each part of the audio. Acting on instructions from the editor these can be reassembled in whatever new order is required. However, the process can be lengthy since DAT playback has to be done in 'real time'. DAT editing is more usually done via a computer. The whole recording is downloaded, manipulated on screen, and played out to a DAT recorder – the audio material remaining throughout in the digital domain.
- *Minidisc* A certain amount of editing can be done on the machine itself. This entails the renumbering of the inserts or tracks in the required numerical order, the unwanted material being placed at the end. The machine then plays them back in this order. Such reordering can be done any number of times without affecting the original sequence. This is also 'non-destructive' editing. It is best not to delete any recording at this stage in case it may be required. Alternatively, the disc can be downloaded to a computer.
- *Quarter-inch tape* Here the tape is actually cut at the edit points, and the unwanted tape physically removed. The process consists of identifying the beginning and the end of the unwanted section by moving the tape slowly by hand past the machine's replay head and listening carefully for the first word to be edited out. This point is lightly marked with a chinagraph pencil – *not* a ballpoint pen to avoid

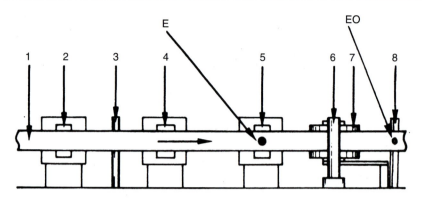

Figure 2.11 Tape machine head assembly. 1. Tape. 2. Erase head. 3. Guide pin. 4. Recording head. 5. Replay head. 6. Capstan. 7. Pinch wheel which holds the tape firmly against the capstan. 8. Guide pin. During editing, the tape is marked on the replay head at point E, or for offset marking at EO

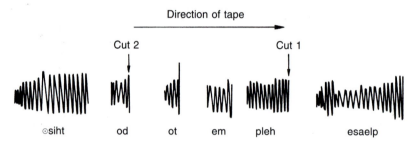

Figure 2.12 Identifying an edit. Since the tape is moving from left to right, the first part of 'please, help me to do this', seen from the front of the heads, is on the right. To edit it to 'please, do this' the cuts are made at the beginning of the appropriate words, not at the end of the preceding ones

damaging the head. The first sound that *is* wanted is similarly identified and marked, these two marks are then cut with a single-sided razor blade in the diagonal slot of an editing block, the unwanted tape removed, and the two ends of the remaining tape joined with approximately 3 cm of adhesive tape.

It is interesting to note that some radio trainers teach the cut editing of tape before computer editing, because it gives a clear idea of what is happening. Digital editing then becomes easier to understand.

An alternative technique, called offset marking, is used where access to the replay head is difficult. This entails marking the tape at a place ahead of the edit point – the distance being taken into account when making the cut on the editing block. The 90° cutting slot on the block is for use when editing music stereo recordings where the cutting of the two tracks must be more precisely at the same place.

- *Audio cassette* Editing is generally done by dubbing or copying it to another format – either computer or reel to reel – before editing. Sometimes only the part or parts required need be copied so quickly

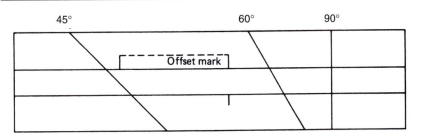

Figure 2.13 Editing block with offset mark. The offset mark is ahead of the cutting slot by the same distance as the guide pin (8) is ahead of the replay head (5). The 60° and 90° slots are used to cut two stereo tracks at more nearly the same place

producing the finished result. This procedure can, of course, also apply to DAT, minidisc, or $\frac{1}{4}$-inch tape, so that only those parts that are wanted are copied to another format. This may well be the most suitable method where long sections of material are to be removed, where the recording levels vary and need adjusting, or where studio links have to be inserted, for example in making a package. In order to maintain a consistent sound level, the different elements are copied through a mixer to make the final recording; this becomes the package ready for transmission (see also Computer – four-track mixing).

- *Retake editing*   A simple studio technique which corrects a mistake without editing involves a retake to cover the error. If a speaker fluffs or makes a mistake, the recording is stopped, and replayed. At the end of the sentence before that in which the mistake was made, the playback is changed to record and the speaker picks up immediately at that point. The offending sentence is erased and with no noticeable gap the correct version takes its place. This procedure, when recording, avoids further editing time for the sake of one or two verbal or operational errors.

CDs, albums, and other discs

Music reproduction employs a wide variety of methods. Some stations have all their music on a central computer system with instant access to any track. Others rely on a library built up over the years of vinyl records – or of cassette tapes. Almost all use CDs in one form or another, either on individual players or through a jukebox system. It is not surprising therefore that studio equipment differs widely from station to station.

Though mostly used as a portable recorder for use on location, the minidisc also features as a playback system in the studio. Since each disc carries 140 minutes of mono programme, 70 in stereo, it is easily capable of handling a one-hour music programme.

Of all the equipment in a studio, it's as well to remember that music gear is particularly sensitive to banging and vibration. CD players don't like being knocked in use, turntables object to being leaned on. Neither do discs of any kind like being left out of their cases or sleeves – they are worth looking after and should not be left lying around gathering dust, cigarette ash, or fingerprints. Compact discs and vinyl records should be held so that the fingers do not touch the playing surface. CDs should be cleaned with a soft non-fluffy cloth, not round and round concentrically since this may affect and damage the tiny pits in the grooves which represent the music. Instead they should be wiped gently radially from the centre outwards.

Given proper care, the compact disc has several key attributes:

- superb reproduction, perfectly suited to FM stereo transmission
- easy automatic track location, with track number indicated
- precision starting, at the beginning of a track or mid-point, e.g. on a vocal
- visual readout of 'time played' and 'time remaining'
- fast forward and fast backward 'audible search' facility to find a particular cue
- long playing time, its one side able to play up to 70 minutes
- able to withstand mishandling, although care is required
- reduced storage requirement.

In playing a disc, most control desks have a 'prefade', 'pre-hear' or 'audition' facility which enables the operator to listen to the track and adjust its volume before setting it up to play on air. This provides the opportunity of checking that it is the right piece of music, but listening only to the beginning may give a false idea of its volume throughout.

Instead of individual CD players, some stations use a form of CD carousel or jukebox holding perhaps 100 discs representing 2000 tracks. This avoids the need for the presenter to handle discs, thereby overcoming several problems. Discs do not get lost, stolen, or mislaid, and the playing surface is kept in perfect condition with no finger marks – the main cause of track skipping. The jukebox, which may not be in the studio but in a separate control room, is accessed through a music selection system, for example a touch-screen computer. The system can be expanded by linking jukeboxes in tandem and is clearly ideal for a station where track availability at any one time is determined centrally as a matter of policy.

A possible argument against remote operation of CDs is that presenters often prefer to feel in control of their programme through physical contact with their discs. The disadvantage here is that the disc is then separated from its inlay notes. However, this information can always be made available from a data store, brought up on a computer screen as required. The computer itself is a further method of storing music for playout since a large-capacity hard disk can store several thousand tracks.

A fully digital station, using only CDs or computer disc as a music source, is wise to keep at least one conventional turntable for vinyl records – even the occasional 78 rpm disc can be brought in by a listener.

Microphones

The good microphone converts acoustic energy into electrical energy very precisely. It reacts quickly to the sudden onset of sound – its transient response; it reacts equally to all levels of pitch – its frequency response; and it operates appropriately to sounds of different loudness – its sensitivity and dynamic response. It should be sensitive to the quietest sounds, yet not so delicate as to be easily broken or susceptible to vibration through its mounting. It should not generate noise of its own. Add to these factors desirable qualities in terms of size, weight, appearance, good handling, ease of use, reliability and low cost, and microphone design becomes a highly specialised scientific art.

To the producer the most useful characteristic of a microphone is probably its directional property. It may be sensitive to sounds arriving from all directions – omni-directional – and such a microphone is useful for location recording and interviewing, audience reaction, and talkback

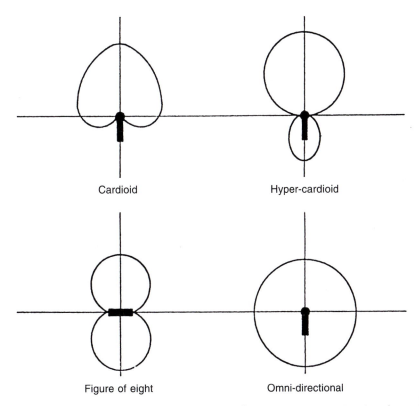

| Cardioid | Hyper-cardioid |

| Figure of eight | Omni-directional |

Figure 2.14 Microphone polar diagrams or directivity patterns. A microphone is sensitive to sounds within its area of pick-up but in selecting one for a particular purpose consideration must also be given to how well it will reject unwanted sounds from another direction

purposes. Alternatively a directional mic is essential in most types of music balance, quiz shows, and where there is any form of public address system.

The choice of mic for a particular job requires some thought and although it may be possible to rely on the expertise of a technician, it will pay a producer to become familiar with the advantages and limitations of each type available. For example, some mics include on/off switches, or a switch to start a portable recorder. Some incorporate an optional bass-cut facility. Some require a mains unit or battery pack, or have a directivity pattern that can be changed while in use. Some operate better out of doors than others, some will make a presenter sound good when working close to it, others just distort. Producers should have views about whether a radio mic is necessary, when clip-on personal mics are appropriate, or if a highly directional 'rifle' mic is required. The more one knows about the right use of the right equipment, the more the technicalities of programme making become properly subservient to the editorial decisions.

Stereo

Simply stated, the stereo microphone gives two electrical outputs instead of one. These relate to sounds arriving from its left and its right. This 'positional information' is carried through the entire system via two transmission channels arriving at the stereo receiver to be heard on left and right loudspeakers. The left channel is generally referred to as the 'A' (red) output and the right channel is the 'B' (green) output. The meter monitoring the electrical levels has two needles – red and green. The signal sent to a mono transmitter – the 'M' signal – is the *combination* of both left and right, i.e. A + B, while the stereo information – the 'S' signal – consists of the *differences* between what is happening on the left and on the right, i.e. 'A – B'. Sometimes a second monitoring meter is available to look at the 'M' and 'S' signals. Again, it has two needles conventionally coloured respectively white and yellow. Vertical columns of LEDs are an alternative way of indicating the signal level. What does the producer need to know about all this?

First, that if a programme is to be carried by both monaural and stereo transmitters some thought has to be given to the question of compatibility. Material designed for stereo can sound pointless in mono, or even technically bad. For example, speech and music together can be distinguished in stereo purely because of their *positional* difference; in mono the same mix may be unacceptable since a difference in *level* is needed. The producer must make the programme for his primary audience – it is unlikely to be totally right for both. It is all too easy to fall in love with the stereo sound in the studio and forget the needs of the mono listener.

Second, that stereo mics are fairly uncommon and it is not necessary

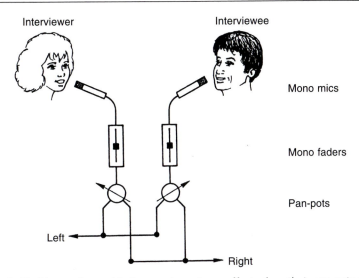

Figure 2.15 Mono mics and faders create a stereo effect when their pan-pots are set to give a left and right placing to each sound source

to use one to generate a stereo signal. Two directional mono mics – or 'co-incident pair' – connected to a stereo mixer in such a way as to simulate left and right signals, for example through 'pan-pots', will give excellent results. This technique is useful in an interview or phone-in when the voices can be given some additional separation for the stereo listener.

Third, that a pan-pot can give a mono source both size and position. For example, a mono recording of a sound effect can be placed across part, or all, of the sound picture. Two mono recordings, for example of rain, can give a convincing stereo picture if one is panned to the left and the other to the right, with some overlap in the middle. When recording music special effects can be obtained by the deliberate mislocation of a particular source – hence the piano 10 metres wide at the front of the orchestra, or the trumpeter whose fanfare flies around the sound stage simply by the twirl of his pan-pot!

Fourth, that stereo working requires a higher standard of studio cleanliness and equipment maintenance. Dusty records, for example, are much more obvious on the air; every speck which affects one channel more than the other shows up as a noticeable stereo click.

And fifth, that working in stereo is a challenge to the producer's creativity. To establish distance and effect movement in something as simple as a station promo gives it impact. To play three records simultaneously – one left, one centre, and one right – as a 'guess the title' competition is intriguing. A spatial round-table discussion which really separates the speakers has a much more live feel than its mono counterpart. The drama – or commercial – in which voices can be made to appear from anywhere, dart around or 'float', literally adds another dimension. For the listener, a stereo station should do more than keep his stereo indicator light on.

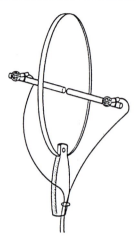

Figure 2.16 A 'dummy head' for binaural stereo. Two omni-directional tie-clip mics, representing the two ears, are fixed some 9 cm from the centre of a Perspex disc 20 cm in diameter

Binaural stereo

For the really adventurous, there is binaural stereo. Using a dummy head, or a head-sized Perspex disc in the 'front to back' plane, two omni-directional microphones are used where the ears would be. The action takes place around this device and the result recorded on two-track stereo in the normal way. For the listener on headphones, the dramatic effect is quite uncanny.

Equipment faults

Studios are complex places: There is a lot to go wrong – a knob or key on the mixer works loose, a CD player gets out of adjustment, the clock is not absolutely accurate, a small indicator light may blow or a headphone lead break. Perhaps it is something physical like a squeaky door or an unsafe mic stand, or operational like unreliable software or a skipping CD track. Whatever the problem it is likely to affect programmes and must be put right. It is up to every studio user to report any fault and the easiest way of doing this is by means of a small notebook permanently available in a conspicuous place. It is the responsibility of the person who discovers the fault to make a note of its symptoms and the time it happened. Every morning the book is routinely checked by a maintenance technician. Intermittent faults are particularly troublesome and may require a measure of detective work. It is in everyone's interest to record any technical incident, however slight, in order to maintain high operational standards.

3

Interviewing

The aim of an interview is to provide, in the interviewee's own words, facts, reasons or opinions on a particular topic so that the listener can form a conclusion as to the validity of what he or she is saying.

The basic approach

It follows from the above definition that the opinions of the interviewer are irrelevant, he should never get drawn into answering a question which the interviewee may put to him – an interview is not a discussion. We are not concerned here with what has been referred to as 'the personality interview' where the interviewer, often the host of a television 'chat show', acts as the grand inquisitor and asks his guests to test their opinions against his own. Within this present definition it is solely the interviewee who must come through and in the interviewer's vocabulary the word 'I' should be absent. Deference is not required but courtesy is; persistence is desirable, harassment not. The interviewer is not there to argue, to agree or disagree. He is not there to comment on the answers he gets. He is there to ask questions. To do this he needs to have done his homework and must be prepared to listen.

The interview is essentially a spontaneous event. Any hint of its being rehearsed damages the interviewee's credibility to the extent of the listener believing the whole thing to be 'fixed'. For this reason, while the topic may be discussed generally beforehand, the actual questions should not be provided in advance. The interview must be what it appears to be – questions and answers for the benefit of the eavesdropping listener. The interviewer is acting on behalf of the listener, asking the questions which the listener would want to ask. More than this, he is asking the questions which his listener would ask if he knew as much about the background to the subject as the interviewer knows. The interview is an opportunity to provide not only what the listener wants to know, but also what he may need to know. At least as far as the interviewing of political figures is concerned, the interview should represent a contribution towards a democratic society; i.e. the proper questioning of people who, because of the office they hold, are accountable to the electorate. It is a valuable element of broadcasting and care should be taken to ensure it is not damaged, least of all by casual abuse in the cult of personality on the part of broadcasters.

Types of interview

For the sake of simplicity three types of interview can be identified, although any one situation may involve all three categories to a greater or lesser extent. These are the informational, the interpretive and the emotional interview.

Obviously, the purpose of the *informational interview* is to impart information to the listener. The sequence in which this is done becomes important if the details are to be clear. There may be considerable discussion beforehand to clarify what information is required and to allow time for the interviewee to recall or check any statistics. Topics for this kind of interview include: the action surrounding a military operation, the events and decisions made at a union meeting, or the proposals contained in the city's newly announced development plan.

The *interpretive interview* has the interviewer supplying the facts and asking the interviewee either to comment on them or to explain them. The aim is to expose his reasoning and allow the listener to make a judgement on his sense of values or priorities. Replies to questions will almost certainly contain statements in justification of a particular course of action which should themselves also be questioned. The interviewer must be well briefed, alert, and attentive to pick up and challenge the opinions expressed. Examples in this category would be a government minister on his reasons for an already published economic policy, why the local council has decided on a particular route for a new road, or views of the clergy on proposals to amend the divorce laws. The essential point is that the interviewer is not asking for the facts of the matter, since these will be generally known; rather he is investigating the interviewee's reaction to the facts. The discussion beforehand may be quite brief, the interviewer outlining the purpose of the interview and the limits of the subject he wishes to pursue. Since the content is reactive, it should on no account be rehearsed in its detail.

The aim of the *emotional interview* is to provide an insight into the interviewee's state of mind so that the listener may better understand what is involved in human terms. Specific examples would be the feelings of relatives of miners trapped underground in a pit accident, the euphoria surrounding the moment of supreme achievement for an athlete or successful entertainer, or the anger felt by people involved in an industrial dispute. It is the strength of feeling present rather than its rationality which is important and clearly the interviewer needs to be at his most sensitive in handling such situations. He will receive acclaim for asking the right question at the right time in order to illuminate a matter of public interest, even when the event itself is tragic. But he is quickly criticised for being too intrusive into private grief. It is in this respect that his manner of asking a question is as important as its content, possibly more so. Another difficulty which faces the interviewer is to reconcile his need to remain an impartial observer with his not appearing indifferent to the suffering. The amount of time taken in preliminary conversation will vary considerably depending on the circumstances. Establishing the necessary relationship may be a lengthy process – but there is a right moment to begin recording and it is important for the interviewer to remain sensitive to this judgement. Such a situation allows little opportunity for retakes.

These different categories of interview are likely to come together in preparing material for a documentary or feature. First the facts, background information or sequence of events; then the interpretation, meaning or implication of the facts; finally their effect on people, a personal reaction to the issue. The *documentary interview* with, for example, a retired politician will take time but should be as absorbing for the interviewer as it will be for the listener. The process of recalling history should surprise, it should throw new light on events and people, and reveal the character of the person. Each interview is different but two principles remain for the interviewer – listen hard, and keep asking 'why?'

Related to the documentary interview, but not concerned necessarily with a single topic, is the style of interviewing which contributes to *oral history*. Every station, national, regional, or local, should assume some responsibility for maintaining an archival record of its area. Not only does it make fascinating material for future programmes but it becomes of value in its own right as it marks the changes which affect every community. The old talking about their childhood, or their parents' values, craftsmen describing their work, children on their expectations of growing up, the unemployed, music makers, shopkeepers. The list is endless. The result is an enriching library of accent, story, humour, nostalgia, and idiosyncrasy. But to capture the voices of people unused to the microphone takes patience and a genuinely perceptive interest in others. It may take time to establish the necessary rapport and to put people at their ease. On the other hand, when talking to the elderly, it is often advisable to start recording as soon as the interviewee recounts any kind of personal memory. They may not understand whether or not the interview has begun and will not be able to repeat what they said with the same freshness. Preparation and research beforehand into personal backgrounds help to recall facts and incidents which the interviewee often regards as too insignificant or commonplace to mention. People generally like talking about themselves and it takes a quick-thinking flexibility to respond appropriately, to know when a conversation should be curtailed, and when moved on. The rewards, however, are considerable. A final point, such recordings need to be well documented – it is one thing to have them in the archives, but quite another to secure their speedy and accurate retrieval.

Preparation before the interview

It is essential for the interviewer to know what he is trying to achieve. Is the interview to establish facts, or to discuss reasons? What are the main points which must be covered? Are there established arguments and counter-arguments to the case? Is there a story to be told? The interviewer must obviously know something of the subject and a briefing from the producer, combined with some research on his own part, is highly desirable. An essential is absolute certainty of any names, dates, figures or other facts used within the questions. It is embarrassing for the expert interviewee to correct even a trifling factual error in a question – it also represents a loss of control, for example:

> 'Why was it only 3 years ago that you began to introduce this new system?'
> 'Well actually it was 5 years ago now.'

It is important, although easily capable of being overlooked, to know exactly who you are talking to:

> 'As the chairman of the company, how do you view the future?'
> 'No, I'm the managing director . . .'

It makes no difference whatsoever to the validity of the question but a lack of basic care undermines the questioner's credibility in the eyes of the interviewee and, even more important, in the ears of the listener.

Having decided what has to be discovered, the interviewer must then structure the questions accordingly. Question technique is dealt with in a later section but it should be remembered that what is actually asked is not necessarily formulated in precise detail beforehand. Such a procedure could easily be inflexible and the interviewer may then feel obliged to ask the list of questions irrespective of the response by the interviewee. Preparation calls for the careful framing of alternative questions – with consideration of the possible responses so that the next line of enquiry can be worked out.

For example, you want to know why a government minister is advocating the closure of coal mines with a large amount of resulting unemployment. If the question put is simply 'Why are you advocating . . .?', the reply is likely to be a stock answer on the need for pits to be economic. Such a response is known by most people so the interview merely repeats the position, it does not carry the issue forward. To move ahead, the interviewer must anticipate and be in a position to put questions that *well-informed people in the industry* are asking – about other markets for coal, the relative operating and capital costs of coal-fired and gas power stations, the cost to the country of unemployed mineworkers, possible relief measures, and so on.

To summarise, an interviewer's normal starting point will be:

1 To obtain sufficient briefing and background information on the subject and the interviewee.
2 To have a detailed knowledge of what the interview should achieve, and at what length.
3 To know what the key questions are.
4 By anticipating likely responses, to have ready a range of supplementary questions.

The pre-interview discussion

The next stage, after the preparatory work, is to discuss the interview with the interviewee. The first few minutes are crucial. Each party is sizing up the other and the interviewer must decide how to proceed.

There is no standard approach: each occasion demands its own. The interviewee may respond to the broadcaster's brisk professionalism or might better appreciate a more sympathetic attitude. He may need to feel important, or the opposite. The interviewee in a totally unfamiliar situation may be so nervous as to be unable to marshal his thoughts properly; his entire language structure and the speed of delivery may be affected. Under stress he may not even be able to listen fully to your questions. The good interviewer will be aware of this and will work hard to enable the interviewee's thinking and personality to emerge. Whatever the circumstances, the interviewer has to get it right, and he has only a little time in which to form his judgements.

The interviewer indicates the subject areas to be covered but he is well advised to let the interviewee do most of the talking. This is an opportunity to confirm some of the facts, and it helps the interviewee to release his own tensions while allowing the interviewer to anticipate any problems of language, coherence or volume.

It is wrong for the interviewer to get drawn into a discussion of the matter, particularly if there is a danger that he might reveal his own personal attitude to the subject. Nor must he adopt a hostile manner or imply criticism. This may be appropriate during the interview but even so it is not the interviewer's job to conduct a judicial enquiry, nor to represent himself as prosecuting counsel, judge and jury.

The interviewer's prime task at this stage is to clarify what the interview is about and to create the degree of rapport which will produce the appropriate information in a logical sequence at the right length. He must obtain the confidence of the interviewee while establishing his means of control. A complex subject needs to be simplified, and distilled for the purposes of say a $2\frac{1}{2}$-minute interview – there must be no technical or specialist jargon, and the intellectual and emotional level must be right for the programme. Above all, the end result should be interesting.

It is common and useful practice to say beforehand what the first question will be, since in a 'live' situation it can help to prevent a total 'freeze' as the red light goes on. If the interview is to be recorded, such a question may serve as a 'dummy' to be edited off later. In any event it helps the interviewee to relax and to feel confident about starting. The interviewer should begin the actual interview with as little technical fuss as possible, the preliminary conversation proceeding into the interview with the minimum of discontinuity.

Question technique

An interview is a conversation with an aim. On the one hand, the interviewer knows what that aim is and he knows something of the subject. On the other, he is placing himself where the listener is and is asking questions in an attempt to discover more. This balance of knowledge and ignorance can be described as 'informed naivety'.

The question type will provide answers of a corresponding type. In their simplest form they are:

1   Who?   asks for fact. Answer – a person.
2   When?   asks for fact. Answer – a time.
3   Where? asks for fact. Answer – a place.
4   What?   asks for fact or an interpretation of fact.
            Answer – a sequence of events.
5   How?   asks for fact or an interpretation of fact.
            Answer – a sequence of events.
6   Which? asks for a choice from a range of options.
7   Why?   asks for opinion or reason for a course of action.

These are the basic 'open' question types on which there are many variations. For example:

'How do you feel about . . .?'
'To what extent do you think that . . .?'

The best of all questions, and incidentally the one asked least, is 'why?' Indeed after an answer it may be unnecessary to ask anything other than 'why is that?' The 'why' question is the most revealing of the interviewee since it leads to an explanation of actions, judgements, motivation and values:

'Why did you decide to . . .?'
'Why do you believe it necessary to . . .?'

It is sometimes said that it is wrong to ask 'closed' questions based on the 'reversed verb':

Are you . . .?
Is it . . .?
Will they . . .?
Do you . . .?

What the interviewer is asking here is for either a confirmation or a denial; the answer to such a question is either yes or no. If this is really what the interviewer is after, then the question structure is a proper one. If, however, it is an attempt to introduce a new topic in the hope that the interviewee will continue to say something other than yes or no, it is an ill-defined question. As such it is likely to lead to the interviewer's loss of control, since it leaves the initiative completely with the interviewee. In this respect the reversed verb question is a poor substitute for a question which is specifically designed to point the interview in the desired direction. The reversed verb form should therefore only be used when a yes/no answer is what is required:

'Will there be a tax increase this year?'
'Are you running for office in the next election?'

Question 'width'

This introduces the concept of how much room for manoeuvre the interviewer is to give the interviewee. Clearly where a yes/no response is being sought, the interviewee is being tied down and there is little room for manoeuvre; the question is very narrow. On the other hand, it is possible to ask a question which is so enormously wide that the interviewee is confused as to what is being asked:

'You've just returned from a study tour of Europe, tell me about it.'

This is not of course a question at all, it is an order. Statements of this kind are made by inexperienced interviewers who think they are being helpful to a nervous interviewee. In fact the reverse is more likely with the interviewee baffled as to where to start.

Another type of question, which again on the face of it seems helpful, is the 'either/or' question:

'Did you introduce this type of engine because there is a new market for it, or because you were working on it anyway?'

The trouble here is that the question 'width' is so narrow that in all probability the answer lies outside it so leaving the interviewee little option but to say 'Well neither, it was partly . . .' Things are seldom so clear cut as to fall exactly into one of two divisions. In any case it is not up to the interviewer to suggest answers; what he wanted to know was:

'Why did you introduce this type of engine?'

Devil's advocate

If the interviewee is to express his own point of view fully and to answer his various critics, it will be necessary for those opposing views to be put to him. This gives him the opportunity of demolishing the arguments to the satisfaction, or otherwise, of the listener. In putting such views the interviewer must be careful not to associate himself with them, nor must he be associated in the listener's mind with the principle of opposition. His role is to present propositions which he knows to have been expressed elsewhere, or the doubts and arguments which he might reasonably expect to be in the listener's mind. In adopting the 'devil's advocate' approach common forms of question are:

'On the other hand it has been said that . . .'
'Some people would argue that . . .'
'How do you react to people who say that . . .'
'What would you say to the argument that . . .'

The first two examples as they stand are not questions but statements and if left as such will bring the interview dangerously close to being a discussion. The interviewer must ensure that the point is put as an objective question.

It has been said in this context that 'you can't play good tennis with a bad opponent'. The way in which broadcasters present counter-arguments needs care, but experienced interviewees welcome it as a means of making their case more easily understood.

Multiple questions

A trap for the inexperienced interviewer, obsessed with the fear that his interviewee will be lacking in response, is to ask two questions at once:

'Why was it that the meeting broke up in disorder, and how will you prevent this happening in future?'

The interviewee presented with two questions may answer the first and then genuinely forget the second, or he may exercise his apparent option to answer whichever one he prefers. In either case there is a loss of control on the part of the interviewer as the initiative passes to the interviewee.

Questions should be kept short and simple. Long rambling circumlocutory questions will get answers in a similar vein; this is the way conversation works. The response tends to reflect the stimulus – this underlines the fact that the interviewer's initial approach will set the tone for the whole interview.

Beware the interviewer who has to clarify his own questions:

'How was it you embarked on such a course of events, I mean what made you decide to do this – after all at the time it wasn't the most obvious thing to do, was it?'

Confusion upon confusion, and yet this kind of muddle can be heard on the air. If the purpose of the question is not clear in the interviewer's mind, it is unlikely to be understood by the interviewee – the listener's confusion is liable to degenerate into indifference and subsequent total disinterest.

Leading questions

Lazy, inexperienced or malicious questioning can appear to put the interviewee in a particular position before he begins:

'Why did you start your business with such shaky finances?'
'How do you justify such a high-handed action?'

It is not up to the interviewer to suggest that finances are shaky or that action is high-handed, unless this is a direct quote of what the interviewee has just said. Given the facts, the listener must be able to determine for himself from what the interviewee says whether the finances were sufficient or whether the action was unnecessarily autocratic. Adjectives which imply value judgements must be a warning signal, for interviewee and listener alike, that all is not quite what it appears to be. Here is an interviewer who has a point to make, and in this respect he may not be properly representing the listener. The questions can still be put in a perfectly acceptable form:

'How much did you start your business with?' (fact)
'At the time did you regard this as enough?' (yes/no)
'How do you view this now?' (judgement)
'What would you say to people who might regard this action as high-handed?' (This is the 'devil's advocate' approach already referred to.)

It is surprising how one is able to ask very direct, personally revealing, 'hard' questions in a perfectly acceptable way by maintaining at the same time a calmly pleasant composure. When a broadcaster is criticised for being over-aggressive, it is his manner rather than what he says which is being questioned. Even persistence can be politely done:

'With respect, the question was *why* this had happened.'

In asking *why* something happened it is not uncommon to get in effect *how* it happened, particularly if the interviewee wishes to be evasive. If he is evasive a second time, this will be obvious to the listener and there is no need for the interviewer to labour the point: it is already made.

Non-questions

Some interviewers delight in making statements instead of asking questions. The danger is that the interview may become a discussion. For example, an answer might be followed by the statement:

'This wouldn't happen normally'

instead of with the question 'Is this normal?'
     A further example is the statement:

'You don't appear to have taken that into account'

instead of the question 'To what extent have you taken that into account?'
     Again, the fault lies in the question not being put in a positive way; the interviewee can respond how he likes, perhaps by asking a question himself, and the interviewer will find it difficult to exercise control over both the subject matter and the timing.
     Occasionally interviewers ask whether they can ask questions:

'Could I ask you if . . .'
'I wonder whether you could say why . . .'

This is unnecessary of course since in the acceptance of the interview there is an agreement to answer questions. There may occasionally be justification for such an approach when dealing with a particularly sensitive area and the interviewer feels the need to proceed gently. This phraseology can be used to indicate that the interviewer recognises the difficulty inherent in the question. Much more likely, however, it is used by accident when the questioner is uncertain as to the direction of the interview and is 'padding' in order to provide himself with more thinking time. Such a device to gain time is likely to give the listener the feeling that his is being wasted.

Non-verbal communication

Throughout the interview the rapport established earlier must continue. This is chiefly done through eye contact and facial expression. Once the interviewer stops looking at his respondent, perhaps for a momentary glance at the equipment or at his notes, there is a danger of losing the thread of the interview. At worst, the interviewee will himself look away, and his thoughts as well as his eyes are then liable to wander. The concentration must be maintained. The eyes of the interviewer will express his interest in what is being said – the interviewer is never bored. He can express surprise, puzzlement or encouragement by nodding his head. In fact, it quickly becomes annoying to the listener to have these reactions in verbal form – 'ah yes', 'mm', 'I see'.

Eye contact is also the most frequent means of controlling the timing of the interview – of indicating that another question needs to be put. It may be necessary to make a gesture with the hand, but generally it is acceptable to butt in with a further question. Of course the interviewer must be courteous and positive to the point of knowing exactly what it is he wants to say. Even the most talkative interviewee has to breathe and the signs of such small pauses should be noted beforehand so that the interviewer can use them effectively.

During the interview

The interviewer must be actively in control of four separate functions – the technical, the direction of the interview, the supplementary question and the timing.

The technical aspects must be constantly monitored. Is the background noise altering so requiring a change to the microphone position? Is the position of the interviewee changing relative to the microphone, or have the voice levels altered? If the interview is being recorded, is the machine continuing to function correctly and the meter or other indicator giving a proper reading?

The aims of the interview must always be kept in mind. Is the subject matter being covered in terms of the key questions decided beforehand? Sometimes it is possible for the interviewer to make a positive decision and change course but in any event he must keep track of where he is going.

The supplementary question. It is vital that the interviewer is not so preoccupied with the next question that he fails to listen to what the interviewee is saying. The ability to listen and to think quickly are essential attributes of the interviewer. He must be able to ask the appropriate follow-up question for clarification of a technicality or piece of jargon, or to question further the reason for a particular answer. Where an answer is being given in an unnecessarily academic or abstract way, the interviewer should ask to have it turned into a factual example.

The *timing* of the interview must be strictly adhered to. This is true whether the interview is to be of half an hour or $2\frac{1}{2}$ minutes. If a short news interview is needed, there is little point in recording for 10 minutes with a view to cutting it to length later. There may be occasions when such a time-consuming process will be unavoidable, even desirable, but the preferred method must be 'to sharpen one's mind beforehand, rather than the cutting process afterwards'. Thus the interviewer when recording keeps a clock running in his head. It stops when it hears an answer which is known to be unusable but continues again on hearing an interesting response. He controls the flow of the material so that the subject is covered as required in the time available. This sense of time is invaluable when it comes to doing a 'live' interview when of course timing is paramount. Such a discipline is basic to the broadcaster's skills.

Winding up

The word 'finally' should only be used once. It may usefully precede the last question as a signal to the interviewee that time is running out and that anything important left unsaid should now be included. Other signals of this nature are words such as:

'Briefly, why . . .'
'In a word, how . . .'
'At its simplest, what . . .'

It is a great help in getting an interviewee to accept the constraint of timing if the interviewer has remembered to tell him beforehand the anticipated duration.

Occasionally an interviewer is tempted to sum up. This should be resisted since it is extremely difficult to do without making some subjective evaluations. It must always be borne in mind that the broadcaster's greatest asset is his objective approach to facts and his impartial attitude to opinion. To go further is to forget the listener, or at least to underestimate the listener's ability to form a conclusion for himself. A properly structured interview should have no need of a summary, much less should it be necessary to impose on the listener a view of what has been said.

If the interview has been in any sense chronological, a final question looking to the future will provide an obvious place to stop. A positive convention as an ending is simply to thank the interviewee for taking part:

'Mr Jones, thank you very much.'

However, an interviewer quickly develops an ear for a good out-cue and it is often sufficient to end with the words of the interviewee, particularly if he has made an amusing or strongly assertive point.

After the interview

The interviewer should feel that it has been an enlightening experience which has provided a contribution to the listener's understanding and appreciation of both the subject and the interviewee. If the interview has been recorded, it should be immediately checked by playing back the last 15 seconds or so. No more, otherwise the interviewee is sure to want to change something and one embarks on a lengthy process of explanation and reassurance. The editorial decision as to the content of the interview as well as the responsibility for its technical quality rests with the interviewer. If for any reason he wishes to re-take parts of a recording, he would be wise to adopt an entirely fresh approach rather than attempt to re-create the original. Without making problems for the later editing, the questions should be differently phrased to avoid an unconscious effort to remember the previous answer. This amounts to having had a full rehearsal and will almost certainly provide a stale end product. The interviewee who is losing track of what is going into the final tape is also liable to remark '. . . and as I've already explained . . .' or '. . . and as we were saying a moment ago . . .'. Such references to material which has been edited out will naturally mystify the listener, possibly breaking his concentration on what is currently being said.

If the interview has been recorded, the interviewee will probably want to know the transmission details. If the material is specific to an already scheduled programme, this information can be given with some confidence. If, however, it is a news piece intended for the next bulletin, it is best not to be too positive lest it be overtaken by a more important story and consequently held over for later use. Tell the interviewee when you hope to broadcast it but if possible avoid total commitment.

Thank the interviewee for his time and trouble and for taking part in the programme. If he has come to the studio, it may be normal to offer his travelling expenses or a fee according to station policy. Irrespective of how the interview has gone, professional courtesy at the closing stage is important. After all, you may want to talk to him again tomorrow.

Style

Evidence suggests that political interviewing reflects the conduct of government nationally. If government and opposition are engaged in hard-hitting debate, with individuals making not only party but personal points, then the media will assume this same style. If, however, opposition views are generally suppressed – or in some cases may not exist at all – then ministers will not expect, and may not allow, any challenge from the media.

Different cultures, different nations, have quite diverse views about authority, even when it has been elected by popular vote. In many places, broadcasters may not interview a government minister without first supplying the questions to be asked. In some it is the minister who provides the questions – a practice hardly in the public interest.

In the West, where authority of any kind is not held in particularly high regard, radio and television interviewing is frequently directly challenging, especially of publicly answerable figures. It should not, however, adopt a superior tone nor become a personal confrontation between interviewer and interviewee. The first sign of this is the interviewer making assertions in order to score points instead of asking questions. The producer must vigorously analyse interviewing style to cut short such tendencies which quickly attract public criticism as the media overreaching itself.

Interviewing 'cold'

One of the more challenging aspects of interviewing is in the long sequence programme, such as a breakfast show, where a number of interviewees have been lined up and brought to the studio. The producer/interviewer has little or no opportunity to prepare with each person beforehand. The situation can be improved either through the use of music to give two or three minutes' thinking time, or by having two presenters interviewing alternately. One essential is to have adequate research notes on each interviewee provided by a programme assistant.

If a presenter is meeting a guest for an immediate interview, the basic information which he or she needs is:

Topic title
The person's name
Their position, role, job, status, etc.
The key issue at stake
This person's view of the issue – with actual quotes if possible
Notes on possible questions or approaches, what other people with
    different views have said.

This kind of interviewing, jumping as it does from subject to subject in a quite unrelated way, calls for great flexibility of mind. Its danger is one of superficiality, where nothing is dealt with in depth or comes to a satisfactory conclusion. '. . . I'm afraid we'll have to leave it there because of time.'

The situation is compounded when the interviewee is at the other end of a telephone and can't be seen. Judging the appropriate style is difficult, rapport between the interviewer and interviewee is slight, the result can feel distant and cold, and the need to press on with the running order can make a hurried ending seem rude as far as the listener is concerned – unaware of studio pressures. All these are factors which the producer of the programme should evaluate. The programme may be fast moving, but is it superficial? It might cover good topics but is it essentially lightweight? Is it brisk at the expense of being brusque?

Location interviews

The businessman in his office, the star in her dressing room, the worker in the factory or out of doors; all are readily accessible with a portable recorder and provide credible programming with atmosphere. Yet each may pose special problems of noise, interruption, and right of access.

To remain legal the producer must observe the rules regarding public and private places. Permission is usually required to interview inside a place of entertainment, business premises, a factory, shop, hospital or school. In this last case it is worth remembering that consent of parents or a guardian should normally be sought before interviewing anyone under the age of 16. Working in a non-public area also means that if a broadcaster is asked to leave it is best to do so – or else run the risk of a charge of harassment or trespass.

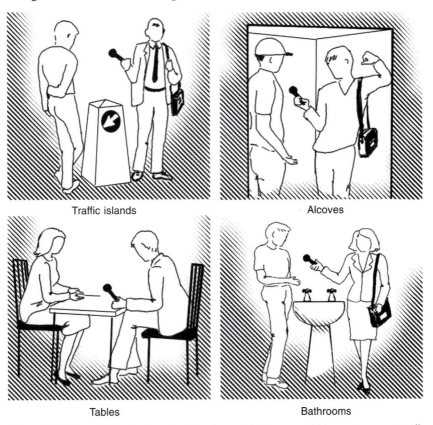

Traffic islands

Alcoves

Tables

Bathrooms

Figure 3.1 Some interviewing situations to avoid. These would be noisy, acoustically poor, or at least asymmetrical, or physically awkward

In any room other than a studio, the acoustics are likely to be poor with too much reflected sound. It is possible to overcome this to an acceptable degree by avoiding hard, smooth surfaces such as windows, desk tops, vinyl floors or plastered walls. A carpeted room with curtains and other furnishings is generally satisfactory, but in unfavourable conditions the best course is to work closer to the microphone, while also reducing the level of input to the recorder.

The same applies to locations with a high level of background noise. Nevertheless the machine shop or aeroplane cockpit need present no insuperable technical difficulty; again the answer is to work closer to the mic and reduce the record level. This will sufficiently discriminate between the foreground speech and the background noise. A greater problem arises where the sounds are intermittent – an aircraft passing overhead, a telephone ringing, or clock striking. At worst these may be so overwhelming as to prevent the interview from being audible but even if this is not the case, sudden noises are a distraction for the listener which a constant level of

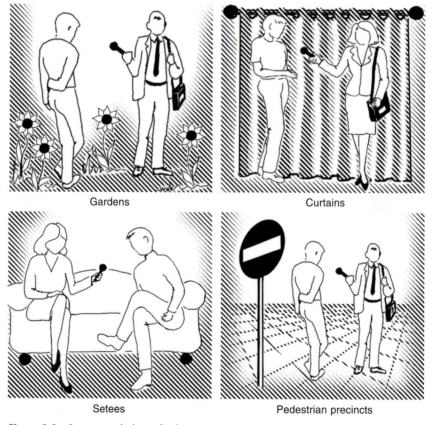

Gardens   Curtains

Setees   Pedestrian precincts

Figure 3.2 Some good places for location interviews. These provide a low or at least a constant background noise, acoustically absorbent surroundings, or a comfortable symmetry

background is not. Background sounds which vary in volume and quality can also represent a considerable problem for later editing – a point interviewers should remember before beginning. The greatest difficulty in this respect arises when an interview has been recorded against a background of music. It is then almost impossible to edit.

Only the very experienced should attempt interviewing using a stereo microphone – and it should then be fixed with a micstand. Even small amounts of movement cause an apparent relocation of the environmental sounds with the consequent disorientation of the listener. Static stereo mics are good for recording location effects, but interviewing almost always calls for a hand-held mono omni-directional mic, or two small clip-on personal mics. In this latter case the stereo recording should be re-mixed on playback with some overlap between the two channels to provide the desired spatial relationship.

It is generally desirable for location interviews to have some acoustic effect or background noise and only practical experience will indicate how to achieve the proper balance with a particular type of microphone. When in doubt, priority should be given to the clarity of the speech.

As with the studio interview, the discussion beforehand is aimed at putting the interviewee at his ease. When outside, using a portable recorder, part of this same process is to show how little equipment is involved. The microphone and machine should be assembled, made ready, and checked during this preliminary conversation. It is important to handle these items in front of the interviewee and not spring the technicalities on him at the last moment. Before starting it is advisable to test the system by 'taking level', i.e. by recording some brief conversation to hear the relative volume of the two voices. If the microphone is to be hand-held, the cable should be looped around the hand, not tightly wrapped round it, so that no part of the cable leading to the recorder is in contact with the body of the microphone. This prevents any movement in the cable from making itself heard as vibration in the microphone. The mic should be out of the eye-line at a point where it can remain virtually stationary throughout. Only in conditions of high background noise should it be necessary to move the mic alternately towards interviewer and interviewee. Even so, this is preferable to the use of an automatic gain control (AGC) on the machine which affects the speaker's voice and background noise together, i.e. it does not discriminate between them as microphone movement does. AGC should therefore be switched 'out', and any noise reduction system, e.g. Dolby, switched 'in'. A satisfactory playback of this trial recording is the final check before beginning the interview.

Some further rules in the use of portable recorders:

1   Where necessary check the ability of the machine to work in unusual conditions, e.g. humidity, electrical radiation or magnetic fields, or at low temperature; or its suitability for specialist functions, such as recording in a coalmine.

Figure 3.3 The microphone cable is formed into two loops. The loop round the finger is kept away from the microphone case to prevent extraneous noises, commonly known as mic rattles, being passed along the cable. The windshield helps to prevent wind noise and voice 'pops'

2   Always check the machine and its microphone before leaving base – record and replay.
3   If there is any doubt about the state of electrical charge of the machine's own cells, take spare batteries.
4   Always use a microphone windshield when recording out of doors.
5   Do not leave a recorder unattended, where it can be seen, even in a locked car.
6   Use the best quality recording materials – discs, tape, or 'chrome' type cassettes.
7   Ensure that rechargeable cells are put back on charge after use.

Interviewing through a translator

News in particular may involve interviewing someone who does not speak your own language. When the interview is 'live', there is no option but to go through an interpreter with the somewhat laborious process of sequential translation – so keep the questions short and simple:

'When were the soldiers here?'
'What has happened to your home?'

'Why did they destroy the village?'
'What happened to your family?'

Depending of course on the circumstances, answers even in another language can communicate powerfully. The translation provides the content of what is said but the replies themselves will define the spirit and strength of feeling in a crisis situation. If the interview is recorded then simultaneous translation is possible through subsequent dubbing and editing. The first question is put, the reply begins and is faded down under the translation. The second question is followed by the translation of the second answer, and so on. It is good to have snatches of the interviewee's voice occasionally, especially in a long interview, as a reminder that we are using an interpreter. What is totally removed in the editing is the translation of the interviewer's questions. Naturally it is best to use a translator's voice similar to that of the interviewee, i.e. a man for a man, or a young girl for a young girl, etc. While it is not always possible to do this in the news context it should be carefully considered for a documentary or feature programme.

4

Being interviewed

Mesmerised perhaps by the 'media moguls' who revel in the cut and thrust of the questioning which appears on the nation's television screens, standing in awe of the interview barons who probe the judgements of the greatest in the land, some people are understandably a little fearful of being interviewed, even by their own local station. They may feel that they are at the mercy of the interviewer and let their minds go blank, hoping for the best. While the previous chapter insists that the overall initiative remains with the interviewer, there are processes which the interviewee must bear in mind – after all, the reason for the broadcast lies within the interviewee, he or she is the expert.

Aims and attitudes

For someone about to be interviewed it is essential to recognise what an interview is and what opportunities it presents, and what it is not.

For example, it is not a confrontation, which it is the interviewer's object to win. An interviewer whose attitude is one of battle, to whom the interviewee represents an opponent who has to be defeated, will almost inevitably alienate the audience. The listener quickly senses such hostility and the probability is that he will side with the underdog. It is easy to cause affront to the listener's sense of fair play and he generally feels that the balance of advantage already lies with the broadcaster. For an interviewer to be unduly aggressive therefore is counter-productive. Neither is an interview a 'point-scoring' operation – this is best left to the debate and the discussion. A would-be interviewee therefore need not feel that he is going to be up against the views of someone else, other than in the indirect 'devil's advocate' form referred to earlier.

On the other hand, an interview is not a platform for the totally free expression of opinion. The assertions made are open to challenge, if not within the interview itself, then possibly in an adjacent 'balancing' interview. The interviewee may not even know about this unless he remembers to ask, but it does illustrate the need for some preparation on his part. He needs particularly to realise that even if nothing positive is said, an impression will certainly have been made. Radio is to do with pictures and the listener will visualise some sort of image to accompany the broadcast. The most useful impression which an interviewee can convey is one of credibility. Only when the listener is prepared to believe in the interviewee, or for that matter any speaker, will he begin to pay much attention to what is being said.

What the interviewee should know

Since it is impossible to interview someone who does not want to be interviewed, it is reasonable to assume that the arrangement is mutually agreed. The broadcaster in contacting a potential interviewee asks whether an interview might take place. The information which the interviewee needs at this point is as follows:

1   What is it to be about? Not the exact questions but the general areas, and the limits of the subject.
2   Is it to be broadcast live or recorded?
3   How long is it to be? Is the broadcast a major programme or a short item? This sets the level at which the subject can be dealt with and helps to guard against the interviewee recording a long interview without his being aware that it must be edited to another length.
4   What is the context? Is the interview part of a wider treatment of the subject with contributions from others or a single item in a news or magazine programme?
5   For what audience? A local station, network use, for syndication?
6   Where? At the studio or elsewhere?
7   When? How long is there for preparation?

No potential interviewee need feel rushed into undertaking an interview and certainly not without establishing the basic information outlined above. Sometimes a fee is paid but this is unlikely in community radio; it is worth asking about.

Shall I be interviewed?

Having obtained an overall picture of what the broadcaster wants, the potential interviewee must ask himself whether he is the right person to be interviewed. Broadcasters will generally approach the person they believe to be most closely involved in the matter but their knowledge of contacts is not infallible. They may go to a company chairman when the PRO would be more appropriate, or they might approach the bishop when one of the lay workers is more informed on the factual detail as opposed to the policy involved.

In deciding whether to be interviewed or not, there may be considerations of possible repercussions at home or at work, the publicity value, or the personal satisfaction of broadcasting. It may also be necessary to ask what the broadcaster would do if there were a refusal to be interviewed. There are at least three courses open to him – to drop the subject altogether; to ask someone else; to broadcast the fact that 'no comment' was forthcoming. This last has the inevitable innuendo that there is something to hide; in this situation the listener should know why a full interview is not possible.

Making time for preparation

The broadcaster may telephone initially and on securing consent to an interview arrive with his portable recorder later that day. Alternatively, he may arrange a studio interview several days in advance. Here there is time for preparation by the interviewee. On the other hand, the interviewer may arrive equipped with his recorder and ask for an interview on the spot – or even telephone and tell the interviewee that he is already on the air! This last technique is bad practice, and besides contravenes the basic right of an interviewee that his telephone conversation will not be broadcast or recorded without his consent. There can be no question of doing this without his knowledge. It should be a standard procedure for all broadcasters that in telephoning a potential interviewee nothing is begun, recorded or transmitted without the interviewee being in possession of all the information outlined earlier. Even so, to be rung up and asked for an interview there and then is asking a great deal and no one should feel obligated to accept this invitation. It is in no one's interest, least of all the listener's, to have an ill-thought-out interview with incomplete replies and factual errors.

The broadcaster may ask for an immediate reaction to a news event and the interviewee may be perfectly prepared, indeed anxious to give it. It is, however, worth making sure that you are fully up to date on the news story before stating your views. It is likely to need thinking through, even briefly; there may be facts to check and other arguments to consider. The broadcaster has his deadline and is not in a position to wait indefinitely, but there may be good reasons why the potential interviewee suggests some slight delay, say ten minutes. This is not to remove the possibility of the immediate interview by telephone, the quick comment, the emotive 'gut' reaction.

Preparing for the interview

Knowing now the subject areas to be discussed, the interviewee needs to crystallise and hold in the front of his mind the *two* most important points he wants to put over. These are the *key points* which he believes should be made – irrespective of the questions he is asked! If he and the interviewer are of a similar mind, these points will be obvious and will occur naturally as the response to questions. If not, the interviewee should include them where he feels it is appropriate.

Recognising the interviewer's role as 'devil's advocate', the wise interviewee will prepare for the possible *counter-arguments* which might be put to him. He should realise that the listener is likely to identify more readily with reasoned argument, based on a capacity to appreciate both sides of a case, than with dogma and bigotry. Accepting the existence of an opposite view and logically explaining why you believe it to be wrong is one of the best ways of sounding convincing on radio.

Whatever is said is enhanced by good *illustration* which underlines the point being made. Drawn from the interviewee's own experience and conjuring up the appropriate pictures, an illustration is a powerful aid to argument. It should be *brief*, *factual*, *recent* and *relevant* – that is, both relevant to the case being put and to the experience of the listener. The intention is that there should be a point of contact, a means by which the listener can identify with the interviewee. For community radio this will almost certainly mean that the illustration is 'local'. For example, in interviewing a police officer about vandalism, a question might be put about unemployment as a possible cause. After answering the question, the interviewee might well illustrate his point with 'For example, only last week we had a case of two boys in the city centre who . . . ' To be effective, illustrations have to be thought about in the interviewee's preparation time.

Check the *essential facts* of the interview – the amount of money committed, the name of the person involved, the tonnage of cargo exported, etc. It is important not to appear too glib, but an interviewee who is in possession of the facts is more likely to gain the respect of the listener.

The interviewee has now prepared his two key points, counter-arguments, illustration and facts, and on meeting the interviewer he confirms with him the aim of the interview, its context and duration.

Nerves

It is not the slightest use telling anyone 'not to be nervous'. Nervousness is an emotional reaction to an unusual situation and as such it is inevitable. Indeed it is desirable in that it causes the adrenalin to flow and improves concentration – with experience it is possible to use such 'red light' tensions constructively. On the other hand, if the interviewee is completely relaxed he may appear to be blasé about the subject and the listener may react against this approach. In practical terms he should listen hard to the interviewer and look at him; eye contact is a great help to concentration.

Making an impression

In aural communication information is carried on two distinct channels –
content (what is said) and *style* (how it is said). They should both be
under the speaker's control and to be fully effective one must reinforce
the other. Due, however, to stress, it is not always easy, for example, to
make a light point lightly. Without intention, it is possible to sound
serious, even urgent and the effect of making a light point in a serious
way is to convey irony. The reverse, that of making light of something
serious, can sound sarcastic. The problem therefore is how to appear
natural in a tense situation. It may be useful to ask yourself 'how should
I come over?'

In discussing the 'images' or impressions which people want to project,
the same epithets invariably occur. Interviewees wish to be seen as friendly,
sincere, human, competent, etc. The following list may be helpful:

1   To be *sincere* – say what you really feel and avoid acting.
2   To be *friendly* – use an ordinary tone of voice and be capable of
    talking with an audible smile. Avoid 'jargon' and specialist language.
3   To appear *human* – use normal conversational language and avoid
    artificial 'airs and graces'. Admit when you do not know the answer.
4   To be *considerate* – demonstrate the capacity to understand views
    other than your own.
5   To be *helpful* – offer useful, constructive practical advice.
6   To appear *competent* – demonstrate an appreciation of the question
    and ensure accuracy of answers. Avoid 'waffle' and 'padding'.

This is of course no different from the ordinary personal contacts
made hundreds of times each day without conscious thought. What is
different for a broadcast interview is that the stress in the situation can
swamp the normal human qualities, leaving the 'colder' professional
ones to dominate. An official concerned to appear competent will all too
easily sound efficient to the point of ruthlessness unless the warmer
human characteristics are consciously allowed to surface.

The most valuable quality is the interviewee's credibility. Only when
the listener believes him as a person will he be prepared to take notice or
act on what he is saying. For this reason, style is initially more important
than content.

Non-answers

The *accidental evasion* of questions may be due to the interviewee genuinely
misunderstanding the question, or the question may have been badly put;
in either case the interview goes off on the wrong tack. When recording
this is easily remedied, but if it happens on the air the listener may be

unable to follow and lose interest, or he may think the interviewee stupid or the interviewer incompetent. One or other of the parties must bring the subject back to its proper logic.

The *deliberately evasive* technique often adopted by the non-answerer is to follow the interviewer's question with another question of his own:

'That certainly comes into it, but I think the real question is whether . . .'

If the new question genuinely progresses the subject, the listener will accept it. If not, he will quickly detect evasion and will expect the interviewer to put the question again. Rightly or wrongly the listener will invariably believe that someone who does not answer has something to hide and is therefore suspect.

There may be genuine reasons why '*No comment*' is an acceptable answer to a question. The facts may not yet be known with sufficient certainty, there may be a legal process pending, a need to honour a guarantee given to a third party, or the answer should properly come from another quarter. It may be that an interviewee legitimately wishes to protect himself from his business competitors – a factor which occurs in the sporting as well as the commercial world.

Nevertheless, the interviewee must be seen to be honest and to say why an answer cannot be given:

'It would be wrong of me to anticipate the report . . .'
'I can't say yet until the enquiry is finished . . .'
'I'm sure you wouldn't expect me to give details, but . . .'

Even if the inability to give a particular answer has been discussed beforehand, an interviewee should still expect to have the question put if it is likely to be in the listener's mind.

In an attempt to slow down the questioning rate or avoid the obvious next question, an interviewee may end his answer with a question back to the interviewer.

'. . . a lot of people are like that, don't you think?'
'. . . so that's what I did; how would you have gone about it?'

It is a golden rule that interviewers never answer questions and the interviewee will get himself ignored for trying to turn an interview into a discussion. The listener, however, may already have gained the impression that the interviewee is unwilling to be interviewed. As with all non-answers this damages his credibility.

The triangle of trust

The whole business of interviewing is founded on trust. It is a three-way structure involving the interviewer, the interviewee, and the listener.

The interviewee trusts the interviewer to keep to the original statement of intent regarding the subject areas and the context of the interview, and also that he will maintain both the spirit and the content of the original in any subsequent editing. The interviewer trusts the interviewee to respond to his questions in an honest attempt to illuminate the subject. The listener trusts the interviewer to be acting fairly in his interests and believes there to be no secret collusion between the interviewer and the interviewee. The interviewee trusts the listener not to misrepresent what he is saying and to understand that within the limitations of time the answers are the truth of the matter.

This 'triangle of trust' is an important constituent not only of the media's credibility but of society's self-respect as a whole. Should one side of the triangle become damaged – for example, listeners no longer trusting broadcasters, interviewees no longer trusting interviewers, or neither having sufficient regard for the listener – there is a danger that the process will be regarded simply as a propaganda exercise. Under these conditions it is no longer possible to distinguish between 'the truth as we see it' and 'what we think you ought to know'. Consequently, the underlying reason for communication begins to disappear, thereby reducing broadcasting's democratic contribution. The fabric of society is affected. This is to take an extreme view, but every time a broadcaster misrepresents, every time an interviewee lies or a listener disbelieves, we have lost something of genuine value.

5
Writing

Writing for radio is the storage of talk. Presentation of a script at the microphone is the retrieval of that talk out of storage. The overall process should give the listener the impression that the broadcaster's <u>talking</u> to him rather than <u>reading</u> at him. It's prepared, of course, but it should <u>sound</u> spontaneous.

This chapter's written in a quite different style from the rest of the book, – it'll contain things which are quite against the rules of 'literary' convention. It may even be difficult for some people to read; the point is, though, that it's in a style which is all right for me to <u>say</u>, – it's <u>spoken</u> English, – a personal talk set out in a way which I would find suitable as a script. Everyone, of course, has to find their own way, and a style such as this for a relaxed 20-minute talk certainly wouldn't do as a voice piece in the news. That would have to be much more tightly written.

I mentioned that this might be difficult to read. There's good reason for this and it's important to realise that writing words on paper is a very crude form of storage. It doesn't give you half what you want in order to make unambiguous sense of it. It doesn't, for example, give you any idea of where the emphasis should go. It's possible to underline certain

words but this isn't the whole story by any means. The
written word provides no indication of the vocal
sounds intended, - the shape of the sentence as it's
said. Yet quite a lot of the meaning of words and
phrases is conveyed in the subtleties of their
inflection. Neither does the page say how it should
sound in terms of speed, or where the pauses should
go. Yet all these qualities help to convey sense -
without them, the words on the paper can be virtually
meaningless - or at least ambiguous. Take this
sentence:

'You mean I have to be there at ten, tomorrow.'
The words stay the same but the precise meaning can
alter eight ways depending where you put the emphasis.
Just try it out loud!

'You mean <u>I</u> have to be there at ten tomorrow?'
'You mean I have to be there, at ten <u>tomorrow</u>'

- and so on.

So <u>writing</u> is only one part of the communication
process, and it isn't completed until the script is
<u>said</u> - and said properly. It's difficult enough to do
well when you're writing something which you're going
to read yourself, it's doubly so when writing for
someone else to read.

Then why do we have a script at all? Wouldn't it be
better just to have notes to 'talk through' at the
microphone? This is possible, but it's not always
advisable - even when the writer is the same person as
the broadcaster. No, there are good reasons for having
a script, like:

- making sure there's as little stress in the
  actual broadcast as possible. The script is a
  'safety-net' - at least you know what you're
  going to say, even though the script won't tell
  you exactly <u>how</u> you're going to say it.

- other points are that a full script makes sure
  that <u>nothing is left out</u>, that it runs to time
  and all the information and arguments are
  presented in the right order. One of the most
  important things about the spoken word is that
  it must be <u>logical</u>. Try telling a joke and
  getting the information in the wrong order! It
  falls flat on its face of course. In print it's
  different because the whole thing is there in
  front of you, – you can look back at something
  again to make sense of it. With 'talk' – it has
  to make sense there and then, or it's likely not
  to make sense at all. And this point depends
  very largely on getting the material in a
  logical order.
- Finally, why we write a script is obviously to
  enable other people to communicate our thoughts
  – to give some permanent form to the otherwise
  very transient nature of speech.

Well, having established the reasons '<u>why</u>', let's
have a look at '<u>what</u>' we're going to write.

The first thing to do, and it applies to any piece
of communication, is to decide what it is you want to
say. What are the points you want to make, and what
sort of impression do you want to leave behind? You
can never be 100 per cent successful in this, but at
least it'll work a good deal better than if you have
no aims at all. You know those times when you're doing
a bit of writing and you stop because you somehow
can't finish the sentence – you've written yourself
into a corner. That's when you cross the sentence out,
put your pen down and say to yourself, 'What am I
trying to say?' It's often a very good question!

So you list out your main points and make sure that

you've got the necessary facts or illustrations to
support them. Then in the cause of logic, you assemble
these ideas and put them in the right order – the order
which makes them most easily understood by your
listener. And you can't do this without having the
listener in mind, – a typical member of your intended
audience. A housewife at home, a child at school, or a
man in his car. It's a great help actually to
visualise your listener as you write, it helps to
prevent it becoming patronising. We often say 'how can
I put this across?' Across, – this implies a horizontal
communication. Not 'how can I put it down?' which is
the patronising attitude – or worse still 'how can I
put it up?' which sounds servile – but 'across' –
someone on the same level as yourself. That's where
the writer should be with regard to his listener.

This approach also stops the writer from suffering
the dread disease of 'difficult' words. We all like to
impress, but if we want to communicate then what's
important is not the impressive word, but the right
word, – the one that's right for our audience, I mean.
After all, if we want to sound as though we're
talking, we'll use the everyday language of our
natural speech rather than the stilted 'officialese'
or jargon that so often appears in print. And if we
want to sound as though we're caring, – caring about
our audience – then the impression we want to make
won't be the broadcaster saying 'aren't I clever!' No,
the point is to express, not impress; but I'll come
back to this business of language in a moment.

A further point about visualising your listener is
to think about broadcasting to one person. Radio is
one of the 'mass media' and it goes to thousands,
millions of people. Yet your message ends up in the

mind of the individual listener. It's wrong to think of radio as a group experience – like an audience in a hall, – or to treat it as a massive public address system reaching out across the housetops and countryside. So avoid phrases like 'listeners may like to know . . .' or 'some of you will have seen . . .' Talking just to one person you'd say, 'you may like to know . . .' and 'If you've seen . . .' It's a communication between you, the broadcaster, and the listener with his own thoughts. So write for the individual, – he'll feel that you're talking just to him and your words'll have much more impact.

So, having decided what we want to say, and in what order – and we know who we're talking to – let's set about the script itself.

Radio is an immensely 'switch-offable' medium, you're talking to a very non-captive audience, so the very first sentence must be interesting. Don't spend a long time 'getting into' the subject, start with an idea that's intriguing, relevant, or at least unusual. And follow it with something which tells me what you're on about, – don't leave me in doubt about that!

'The first sentence must interest, the second must inform.' It's an over-simplified 'rule of thumb' but it's the right idea. Now you go through your list of points linking them together in a logical way, threading them in a sequence like beads on a string. And always make it quite clear what you're doing.

Let me explain. If you get to a difficult point, you might include the phrase, 'how can I explain that?' This is a signal that you're about to digress into an amplification of it. Or you might finish a point and join it on to the next one by saying, 'Let's go on from there to see how this works in practice'.

This is a clear indication to your audience that
you're introducing a new topic, - in this case turning
from theory to its practical application. Such
'indicators' are generally called 'signposts', and
there are plenty in this script, - the trouble is, in
the written word they can be rather tedious, since on
the printed page they normally appear much more
economically as paragraph headings. When spoken
though, they generally sound all right, indeed without
'signposts', it's easy for the listener to lose your
train of thought.

   Now, I said I'd come back to the actual use of
language, and the point I want to make is this; that
the overall style of the broadcast talk should be
conversational. I don't mean it should be sloppy or
casual - some conversation can be quite formal; and in
radio, writing 'the news' generally adopts a more
'careful' approach. But writing the links for a
magazine programme, or compiling the weather forecast,
a religious talk, book review - or any piece of spoken
communication - it should sound colloquial, - like
someone talking. This means that 'it is' becomes
'it's'; 'I would' sounds better as 'I'd'. You have
splendid words like 'this'll' and 'shouldn't've'. They
look terrible on the printed page because we're not
used to seeing them, but listen to people talking, -
they're the stuff of living language, - and good
language at that.

   There's a golden rule for getting the spoken word on
to paper - and that's simply to speak it out loud as
you go along, - and to write down what you hear. Don't
write it in your head, but from the sounds you make as
you talk. Once it's on the page you can always change
it, rearrange and polish it, but the basis of your

script will be a spoken language – much easier to put across in the studio. How many times have you heard someone say – 'as soon as I said it, I knew it was wrong'? Well then, as a broadcaster, don't let anything go down on to paper until you've actually heard it, and known it to be right.

Something else happens when you talk your sentences out loud, – they get shorter. And they're not necessarily the kind of sentences we were taught to write at school. But they have colour and life, and meaning - if you let them. Short sentences are easier to read, and easier for the listener to understand. You get away from the complexities of relative clauses, – this sort of thing:

Jim, who is just about to leave the school where he's been for five years, which included a time as head boy, is looking for a job.

That's not much good on the air, and the three thoughts are much better written as three sentences:

Jim has been at school for five years. This includes a time as head boy. He's now about to leave and is looking for a job.

A further point about saying your script out loud as you write it is that it helps to avoid the tongue twister or the unintentional meaning. The following examples are quite genuine pieces of broadcast script. They may look all right on paper, but try them out loud:

- Six Swedish fishing vessels sailed into the Skagerrak.
- This morning our reporter spoke to Mr . . . on the golf course, as he played a round with his wife.
- The Union said the report was wrong.
- At first, supplies of the new car would be restricted to the home market.

I'm still wondering about this last one. I think it means that the new car would only be sold at home. But couldn't it also mean that supplies to the home market would be restricted and that production was all going for export? It depends on the emphasis given to the word 'restricted'. The sentence itself is unclear.

Here's another example I heard in a major news bulletin. It was after a shooting incident in the street. 'The man was found lying on the pavement by his wife'. Now, what picture does that give you? Does it mean he was lying on the pavement beside his wife? Or should it have been – 'His wife found him lying on the pavement'? The point is that your meaning should be unambiguous – even to someone who's not totally concentrating on what you're saying.

Punctuation should help the reader extract the sense from the writing, but it's a very flimsy device when the whole meaning depends on a comma or two. Take my third example – The Union said the report was wrong. As it stands, it's the report that was wrong and the Union that's saying so, – the Union said the report was wrong. Put in two commas and the situation is exactly reversed, – the Union, said the report, was wrong. Look hard at any sentence containing a dependent clause or bracketed phrase, – is there an easier, a more natural way of saying it? In punctuating a script, I'm very fond of the 'comma dash', – you may have noticed! This joins phrases together yet shows me that there ought to be a small pause. Some people's scripts are covered with all sorts of marks, – arrows to indicate inflection, musical notation, even directions like 'smile', – everything designed to help the reader re-create the way he originally wanted it to sound.

So much for punctuation. The really important thing

is that when you make your points, illustrate them
with parables, and anecdotes, – tell stories, re-live
events. And remember that radio is an immensely visual
medium, – paint pictures, – and yes, in colour, and
appeal to all the senses, – the sense of smell, the
sense of touch so we can all appreciate the roughly
dimpled skin of this brilliant fresh orange that I
have here in my hand, – and the pungent smell and
squirting juice as I cut into it. Ah – and – mm – the
succulent sweetness of this first bite. Did you see
something of that? I hope so for it is the stuff of
radio, – for a talk, or a live commentary.

Some points now on the mechanics of a script. If at
all possible it should be typed, double or even triple
spaced to make it easier to read – a large typeface
rather than a small one, – and leave clear margins on
both sides. This enables alterations and additions to
be made without obliterating the part you want. If a
script has to be written in longhand, make sure it's
totally clear for the reader. Some people put
handwritten scripts all in capital letters, but this
reduces the amount of information on the page in terms
of how you read it, – there's less warning, for example,
about the start of a new sentence, or a 'proper' name.

The script used in the studio should be on good
quality paper, – it's much quieter to handle; and to
avoid unnecessary turning of pages, type only on one
side of the paper.

The page itself should be set out with clear
paragraphs indicating the separate thoughts. A
sentence shouldn't run over to the next page but each
sheet end with a full stop. Indeed, sometimes it may
be desirable not to split an important phrase across
two lines of script, – after all you wouldn't do it if
you were writing figures.

On the subject of numbers, there are three schools
of thought, – writing them in figures, writing them in
words, or putting it both ways. For example:

The Ministry of Defence expenditure on this item
would amount to one and a quarter million pounds.

The cost of the new engine, only came to $380.

Twenty-seven (27) people were injured in the
accident.

The only rule is that the meaning should be clear. The
reader can always cross out the form he doesn't want.

Now, what about the speed and timing? Our reading
speed varies, but a rate of 160 to 180 words a minute
would be normal for a newscast. To give a quick
reckoning on the time a script will take, a single
typed line is 3 to 4 seconds, and a double-spaced page
of A4, – 27 lines, say 270 words, about $1\frac{1}{2}$ minutes. A
certain amount of precision is needed here, a thirty
second voice report means about 85 words, a $2\frac{1}{2}$ minute
piece for the breakfast programme will be 450. It's a
very great art, and a discipline, to get what you want
to say to fit neatly into the time allowed, but the
clock is a hard taskmaster.

Now, on finishing. We started with an interesting
sentence, and it's often a good idea in a general talk
to end with a reference back to that same thought. It
reinforces the point and can act as a 'trigger' for
the later recall of what you said. Of course, if you
want to leave your listener with a specific thought,
or motivate him to a particular action, then these
points must come right at the end. What I'm trying to
say here is that there must be an end, – not a sudden
stop, or a drifting away, – but a clear 'rounding off'.
A resumé perhaps, or a provocative question to
stimulate the listener to further thought. Openings

and closings – without doubt the most difficult part of
any broadcast, – but the final word is often how you'll
be remembered.

And if, in between the beginning and the end, you've
avoided the kind of convoluted literary prose which so
often comes out when we put pen to paper, but instead
have used words in a style which has the texture of a
living, spoken language, – and if the listener can
understand the language, he may understand the
content. He may even have come closer to understanding
you. In fact, you'll have communicated.

So to summarise on what's often called 'writing for
the ear'. Here are my twelve points:
  - decide what you want to say.
  - list your points in a logical order.
  - make sure the opening is interesting and
    informative.
  - write for the individual listener, – visualise
    him or her as you write.
  - speak out loud what you want to say, then write
    it down.
  - use 'signposts' to explain the structure of your
    talk.
  - paint pictures, tell stories, and appeal to all
    the senses.
  - use ordinary conversational language.
  - write in short sentences or phrases.
  - use punctuation to aid clarity for the reader.
  - type the script, double spaced, wide margins
    with clear paragraphs.
  - and when in doubt, keep it simple, – remember,
    the idea is to express, not impress.

Having got our script on to paper, I'll be dealing
later on with Presentation – that's the art of getting
it off again!

6

Cue material

The paperwork which accompanies a recorded interview or other item such as a programme insert has two quite separate functions. The first is to provide the studio staff with information. The second is to introduce the item for the listener so that it makes sense in its context.

The general rules concerning cue material apply equally to the links between items in a sequence programme and to the introductory announcements given to whole programmes.

Information for the broadcaster

Before a recorded programme or item can reach the air, the producer, presenter or studio operational staff require certain information about it. The way these details are laid out on a single page is standard and a sheet of cue material is recognisable as such in most radio stations round the world. It must indicate the following:

1   Name of the radio station.
2   Title of the programme.
3   Date of the intended transmission.
4   Reference title of the piece, a 'catch line'.
5   Suggested on-air introduction.
6   The 'in' and 'out' cues of the programme as they appear on the recording.
7   Precise duration of the material.
8   Suggested on-air 'back-announcement'.
9   Any additional details of a technical or programme nature – editing requirements, tape speed, mono/stereo, etc.

The cue material sheet may also include a note of payment to be made to the contributor. Thus all the information about the programme or item is drawn together on one page, as in the example below.

If there is a recording number this should appear at the bottom of the cue sheet. However, numbering systems can be time consuming to administer and no system should be introduced unless it saves effort and is actually useful. There is sufficient information on the example given here for the recording to be easily identified and most radio stations do without numbers, especially for inserts.

The catch-line should be repeated on the minidisc, cassette, or spool itself – written on a small adhesive label rather than on the plastic.

The producer can often make his decisions based on the cue sheet information – whether in fact to use the item, or where to place it in the programme. If he knows the contributor well and is short of time, he can sometimes use the material without hearing it first. This is not a recommended procedure, and the overall responsibility for the programme still rests with the producer.

1	***Radio Norfolk – Cue Sheet***
2	**PROGRAMME:** Countywide
3	**DATE:** 26th October
4	**CATCHLINE:** Boat appeal
5	**OPENING ANNO:**
	Norfolk's best known charity is launching a major new appeal to help people with disabilities. The Friends of Freedom Group which already runs outings on its own boat is hoping to raise £500,000 to provide a new boat and build an adventure club.
	Our reporter Julie Thompson asked the volunteer skipper of the MV Freedom, Peter Campion, what he hoped to do as a result of the appeal.
6	**CUE IN:** (Water FX fade up) One of the first things will be . . .
	CUE OUT: . . . at the beginning of next year. (boat FX fade 3″)
7	**DURATION:** 2′47″
8	**BACK ANNO:** Peter Campion from the Friends of Freedom Group. And that appeal opens on Saturday.
	PAYMENT
9	Ready for transmission Thompson
	£. . . . . .

The studio operator also has all the information he needs to ensure the recording gets safely on the air. Prior to transmission it is checked to 'take level' and to confirm the 'in' cue.

The studio presenter has a clear indication of how to introduce the item. He may have to alter words to suit his own style, but the good cue writer will write in a way which fits the specific programme.

In the example here the interviewer's first question has been removed from the recording and transferred to the introduction; the insert begins with the first answer. This style is a very common form of cue but it is only one of many and it should not be over-used, particularly within a single programme. It is easy for cue material to become mechanical – to 'write in a rut' – it is important to search for fresh approaches, and some are given in later examples.

Information for the listener

There is an art to writing good cue material. The piece of writing which introduces an item has three functions for the listener. It must be interesting, act as a 'signpost' and be informative.

The information must be *interesting*. The first sentence should contain some point to which the listener can relate. It should be written in response to likely questions: 'What is the purpose of this interview?' 'Why am I broadcasting this piece?' Having found the most interesting facet – the 'angle' most relevant to the listener – the writer starts from that. But more than this, it should be relevant to as many listeners as possible. It should not be written so as to exclude people. To take a local example:

> *Not:* There's a big hole in the road at the corner of Campbell Street and the Broadway.
>
> *But:* They are digging up the road again. Everywhere you go traffic is diverted by a hole with workmen in it. What's going on? For example, at the corner of . . .

The first intro will only interest people who drive down Campbell Street. The second is aimed at all road users, pedestrians too. To draw people in, write from the general to the particular. Another approach is to ask the occasional question as an attempt to involve the listener in a subconscious response.

Cue material is a *signpost* and should make a promise about what is to follow. Having gained the listener's interest, it is then important to satisfy his expectations.

The introduction must be *informative*. One purpose of an introduction is to provide the context within which the item may be properly understood. There may have to be:

1   A summary of the events leading up to or surrounding the story.

2   An indication of why the particular interviewee was chosen.
3   Additional facts to help the listener's understanding. It may be necessary
      to clarify technical terms and jargon, or to explain any background
      noise or sounds which would otherwise distract the listener.
4   The name of the interviewer/reporter.

This last piece of information, generally the last words of the cue, can
become a dreadful cliché:

'Our reporter, John Benson, has the details.'

This is a common introduction to reporter packages and wraps and if
overdone becomes as boring as the 'and he asked her' introductions
noted above for interviews. Cue writing therefore needs a fertile imagination
in order to avoid predictable repetition.

Unless the interview or voice piece is very short, say less than a
minute, it will be necessary to repeat after an interview the information
about the interviewee. There is a high probability that the listener is not
wholly committed to the programme and heard the introduction only
superficially, despite a compellingly interesting opening line. The listener's
full attention frequently becomes engaged only during the interview itself.
Having become fascinated or outraged, it is afterwards that he wants to
know the name of the interviewee and their qualification for speaking.

Radio is prone to fashion and there has been a tendency for the 'back-
announcement' to be omitted. It is said that it slows the programme
down. It is true that a 'backward-pointing' signpost may have such an
effect, whereas introductions help to drive the programme forward.
However, the argument in favour of a back-announcement puts listener
information above programme pace. It also helps to give the impression
that the presenter has been listening. Without some reference to the
interview, the presenter who simply continues with the next item can
sound discourteous. Broadcasters should remember that to the listener,
pre-recorded items are people rather than discs or computer playouts.
They should therefore be referred to as if they had actually taken part in
the programme.

The practice of omitting the back-announcement is probably an example
of radio being influenced by television. In vision, it is possible to
superimpose on an interview a caption giving the name and qualification
of the interviewee. This can happen at any time throughout a piece and
makes a back-announcement unnecessary. The two channels of television
information, sound and vision, can be used simultaneously for different
purposes. This is not the case with radio where a statement of the
interviewee's name is often the simplest and most logical way of
're-informing' the listener.

Two further examples will illustrate the functions of cue material – that
is, to obtain the listener's interest, provide context, explain background
noise, clarify technicalities and to 'back-announce'.

Example one

```
ANNCT:      The strike at Abbots Electrical is over.
            Involving 45 assembly workers and lasting nearly
            two weeks, the strike has meant a loss of
            production worth over £100,000.

            The dispute began when three men were sacked for
            what the management called, "persistent lateness
            affecting the productivity of other workers".

            The Union objected, saying that the men were
            being "unduly victimised".

            Two of the men have since been reinstated. Is
            such a stoppage worth it? On the now busy shop
            floor, our reporter spoke to the Union
            representative, Joe Frimley.

CUE IN:     (noise 3") No stoppage is ever . . .

CUE OUT:    . . . making up for lost time. (noise 2")

                                          Duration 2' 08"

ANNCT:      Joe Frimley, the Union representative at Abbots
            Electrical.
```

When a recording is made against background noise, it is useful to begin the piece with two or three seconds of the sound alone. The insert can be started before the cue and faded up under speech so that its words begin neatly after the introduction. Similarly at the end of the insert, the background noise is faded down behind the presenter's back-announcement. Such a technique is preferable to the jarring effect of 'cutting' on to noise.

In the interests of fairness and objectivity, such an item would invariably need to be followed and 'balanced' by a management view of the situation.

Example two

```
ANNCT:      Space research and your kitchen sink. It seems
            an unlikely combination but the same advanced
            technology which put man on the moon has also
            helped with the household chores.

            For example,  the non-stick frying pan uses a
            chemical called Poly-tetra-fluoro-ethylene
            (Pron: Poli-tetra-flŏorō-éthileen). Fortunately
            it's called PTFE for short. Used now for kitchen
            pans, this PTFE was developed for the coating of
            hardware out in space.
```

```
                    Dr John Hewson of the National Research Council
                    explains.

CUE IN:     We've known about PTFE for some time . . .

CUE OUT:    . . . always looking to the future.

                                        Duration 3' 17"

ANNCT:      Dr Hewson
```

In this case the cue material had the job of explaining what PTFE is. It was referred to in the interview without clarification, so the listener has to be prepared for the term. Cue material is a great way of solving any problem of this sort in a pre-recorded item.

The last name in the cue is generally – but not always – the first voice on the insert. To cue the 'wrong' person is confusing. For example: 'Our reporter Bill West has been finding out how the building work is going.' The voice that follows is assumed to be that of Bill West; if it is actually the site manager it may be some time before the listener realises the fact.

In addition, therefore, to the several functions of cue material, the writer seeks both variety and a lack of ambiguity. Cues and links that are well thought-out will make a real difference in lifting a programme above the rest. It may take time, but it will be time well spent.

7

News – policy and practice

The best short definition of news is 'that which is new, interesting, and true'. 'New' in that it is an account of events which the listener has not heard before – or an update of a story familiar to him. 'Interesting', in the sense of the material being relevant, or affecting him in some way. 'True', because the story as told is factually correct.

It is a useful definition not only because it is a reminder of three crucial aspects of a credible news service but because it leads to a consideration of its own omissions. If all news is to be really 'new' a story will be broadcast only once. Yet there is an obvious obligation to ensure that it is received by the widest possible audience. At what stage then can the news producer update a story, assuming that the listener already has the basic information? What do we mean by 'interesting' when we speak not of an individual but of a large, diversified group with a whole range of interests? Do we simply mean 'important'? In any case, how does the broadcaster balance the short-term interest with the long? And as for the *whole* truth – there simply is not time. So how should we decide out of all the important and interesting events which confront us what to leave out? And concerning what is included, how much of the context should be given in order to give an event its proper perspective? And to what extent is it possible to do this without indicating a particular point of view? And if the broadcaster is to remain impartial, do we mean under *all* conditions?

These are some of the questions involved in the editorial judgement of news. To begin with we need to consider not the practical solutions but the criteria by which possible answers may be assessed.

Starting with the listener, what does he expect to hear? Certainly in a true democracy he has a general right to know and discuss what is going on around him. There will be limitations, defined and maintained by law – matters of national security, confidences of a business or private nature, to which the public does not have rightful access. But these reasons can be used to cloak the genuine interest of the individual. Caught in such a conflict, the broadcaster is faced with a moral problem – the not-unfamiliar one of deciding the greater good between upholding the law and championing the rights and freedom of the individual. At such times, those involved in public responsibility should consider two separate propositions:

1   Broadcasters are not elected: they are not the government and as such are not in a position to take decisions affecting the interest of the State. If they go against the practice of the law they do so as private citizens, with no special privileges because they have access to a radio station.
2   Associated with the public right to know is the private right not to divulge. A society which professes individual freedom does not compel or allow the media to extract that which a person wishes lawfully to keep to himself.

Thus the listener has a right to be informed; but although the constraints may be few and the breaches of it comparatively rare, the right is not total. Every broadcaster must know where he stands and on what basis his lines of editorial demarcation are drawn.

Objectivity

Some declare it to be impossible, that we are inevitably creatures of our own age and environment, seeing the world through the filters of a particular time and culture. In this sense only God is truly objective. But broadcasters must be concerned with truth – even when quite different perceptions and beliefs are held to be true. Objectivity here means recounting these truths accurately and within their own context, even when they conflict with our own personal values. The difficulty is that professional news judgements must, in the end, rely on personal decisions. This is why the question of individual motivation is so important: *why* do I wish to cover this story in this way? – to tell the truth, or to make a point?

In the case of the BBC, the basis of news and current affairs broadcasting has always been – and still is – first, to separate for the listener the reporting of events (news) from the discussion of issues and comment (current affairs), and second, to give both sides of an argument. This is best achieved from the position of being independent of both. Of course there are journalists who see broadcasting as a means of indulging their own attempts at public manipulation, just as there are governments which see news purely in terms of propaganda for their own cause. But people sheltered from unpalatable truths cannot decide, and cannot grow. Of course no government wants, for example, to publicise corruption in high places, but suppose it is there and the broadcasters know that, should it not be investigated and exposed? The political history of the last 200 years shows that if the media does not do it, eventually the people will. Arising from the broadcaster's privileged position as the custodian of this form of public debate, the role of a radio service, even one under government or commercial control, is to allow expression to the various components of controversy but not to engage itself in the argument nor to lend its support to a particular view.[1] For example, where news is written by a government office, there is always the emphasis on supporting its own achievements – 'The government has performed another miracle today . . .' However, objectivity requires a news channel always to distance itself, even slightly, from its sources. This is better as – 'The government says it has performed another miracle today . . .' The difficulty stems from the fact that many governments of new nations see their essential task as nation building, so they believe they have to say how good everything is. Their broadcasting is not about challenging the mediocre or wrong, so you can't publish anything detrimental if you want to keep your job. Nevertheless ways should be found of expressing truth so that broadcasting is fully credible.

What the producer must not do is to introduce a partiality as a result of his own conscious but undisclosed personal convictions and motivation, even for the best of reasons. He must avoid decisions based on his own religious, political or commercial views – since this is putting himself before his listener. The impartiality of chairmanship is an ideal to which the producer must adhere; any bias will seriously damage his credibility for honest reporting. Yet in a world when one man's 'terrorist' is another man's 'freedom fighter', the very language we use in imparting the facts is itself a matter of dispute and allegiance. In this example one learns to use other, more neutral, words like 'guerrilla', and 'gunman'.

Objectivity becomes more difficult and more crucial as society becomes less ordered in its deliberations and more torn with its own divisions. This is something which many countries have witnessed in recent years. The crumbling of an established code of behaviour alters the precepts for making decisions – it may be possible to act impartially in a discussion on, say, the permissive society, but the rest of the station's output is likely to indicate clearly the broadcaster's viewpoint. What is the meaning of impartiality when covering a complex industrial dispute involving official and unofficial representatives, breakaway groups, vocally militant individuals and separate employers' and government views and solutions?

Even more difficult situations are those such as Northern Ireland where there has been a 'limited' civil war. Do we give equal time for those who would uproot society – for those who oppose the rule of law? These are not easy questions since there is a limit to the extent to which anyone may be impartial. When one's own country is involved in armed conflict it is probably not possible or even desirable to be neutral – but one must,

---

[1] This policy of impartiality is by no means universal and in some countries stations are encouraged to take an editorial line. In Britain, radio was for many years the monopoly of the BBC, and had to be as objective as possible. Where there are several broadcasting sources, each may develop its own attitude to political and other controversial issues and like a newspaper attempt to sway public opinion.

as far as possible, remain truthful. While society may be divided and changing in its regard for what is right and wrong, it is less so in its more fundamental approach to good and evil. No public medium of communication can function properly and without critical dissent unless society is agreed within itself on what is lawful and unlawful. It is possible to be impartial in a peaceful discussion on attempts to bring about changes in the existing law, but such impartiality is not possible in reporting attempts to overthrow it by force. One can be objective in reporting the activities of the man with the gun, but not in deciding whether to propagate his views.

A former Director General of the BBC, Sir Hugh Greene, said in the 1960s, 'I do not mean to imply that a broadcasting system should be neutral in clear issues of right and wrong, even though it should be between Right and Left. I should not for a moment admit that a man who wanted to speak in favour of racial intolerance has the same rights as a man who wanted to condemn it. There are some questions on which one should not be impartial.'

There are those who disagree that race relations is a proper area for showing partiality just as there are those who oppose the underlying acceptance of the Christian faith as a basis for conducting public affairs. This is not an abstract or purely academic issue, it is one which constantly faces the individual producer. He must decide whether it is in the public interest to give voice to those who would challenge the very system of democracy which enables him to provide that freedom of expression. On the one hand, to give them a wider currency may be interpreted as a form of public endorsement, on the other, to expose them for what they are may result in their total censure. What is important is the maintenance of the freedom to exercise that choice, and ultimately to be accountable for it to an elected authority. Sir Geoffrey Cox, former Chief Executive of ITN (Independent Television News), has said of the broadcaster's function: 'It is not his duty, or his right, to editorialise on the question of democracy, to advocate its virtues or attack its detractors. But he has a firm duty to see that society is not endangered either because it is inadequately informed, or because the crucial issues of the day have not been so probed and debated as to establish their truth. A good broadcast news service is essential to the functioning of democracy. It is as necessary to the political health of society as a good water supply is to its physical health.'

Democracy cannot be exercised within a society unless its individual members are given a choice on which to make their own moral, political, and social decisions. That choice does not exist unless the alternatives are presented in an atmosphere of free discussion. This in turn cannot exist without freedom – under the law – of the press and broadcasting. The key to objectivity lies in the avoidance of secret motivation and the broadcaster's willingness to be part of the total freedom of discussion – to know that even his editorial judgement, the very basis of his programme making, is open to challenge. Keep the listener informed about what you are doing and why you are doing it – that is the public interest.

News values

From all the events and stories of the day how does the broadcaster decide what to include in the news bulletin? A decision to cover, or not to cover, a particular story may itself be construed as bias. The producer's initial selection of an item on the basis of it being worthy of coverage is often referred to as 'the media's agenda-setting function'. The extent to which a producer allows his own judgement to select the items for broadcast is a subject for much debate. People will discuss what they hear on the radio and are less likely to be concerned with topics not already given wide currency. So is a radio station's judgement as to what is significant worth having? If so, the process of selection, the reasons for rejection, and the weight accorded to each story (treatment, bulletin order and duration) are matters which deserve the utmost care.

There is sufficient evidence to support the significance of the *primacy* and *recency* effects in communication. This means that items presented at the beginning of a bulletin have greater influence than those coming later – also that the final statements exert an inordinate bearing on the total impact – probably because they are more easily recalled. These principles are made much use of in debates and trials but clearly apply also to bulletins, interviews, and discussions. Who speaks first and who is allowed the last word is often a matter of some contention.

The broadcaster's power to select the issues to be debated – and their order of presentation – represents a considerable responsibility. Yet given a list of news stories a group of editors will each arrive at broadly similar running orders for a news bulletin designed for a particular audience. Are there any objective criteria on this matter of news *values*?

The first consideration is to produce a news package suited to the style of the programme in which is it broadcast, answering the question, 'What will my kind of listener be interested in?' A five-minute bulletin can be a world view of twenty items, superficial but wide ranging; or it can be a more detailed coverage of four or five major stories. Both have their place, the first to set the scene at the beginning of the day, the second to highlight and update the development of certain stories as the day proceeds. The important point is that the shape and style of a bulletin should be matters of design and not of chance. Unlike a newspaper with its ability to vary the type size, radio can only emphasise the importance of a subject by its placing and treatment. A typical five-minute bulletin may consist of eight or nine items, the first two or three stories dealt with at one minute's length, the remainder decreasing to thirty seconds each or less. The point was made earlier that compared with a newspaper this represents a very severe limitation on *total* coverage.

Having decided the number and the length of items, the news producer has to select what is important as opposed to what is of passing interest. When short of time it is easier to gain the interest of the listener with an item on the latest scandal than with one on the state of the economy. The second item is more significant for everyone in the long term, but requires more contextual information. The producer must not be put off by such difficulties, for it is the temptation of the easy option which leads to some justification in the charge that 'the media tends to trivialise'. An effect of the policy that news must always be available at a moment's notice is that stories of long-term significance do not find a place in the bulletin. It is after all easier to report the blowing up of an aircraft than the development of one.

A second criterion for selection is to favour items to do with people rather than things. The threat of an industrial dispute affecting hundreds of jobs will rate higher than a world record price paid for a painting. 'How could this event affect my listener?' is a reasonable question to ask. For the listener to a local station in Britain, fifty deaths following an outbreak of typhoid in Hong Kong would be probably regarded with less significance than a road accident in his own area in which no one was hurt. But should it? Particularly in local radio there is a tendency to run a story because of its association with mayhem and disaster rather than its relevance. A preoccupation with house fires and traffic accidents, otherwise referred to as 'ambulance chasing', is to be discouraged.

News values resolve themselves into what is of interest to, or affects, the listener. Essentially they are determined by what is:

- Important – events and decisions that affect the world, the nation, the community, and therefore me.
- Contentious – an election, war, court case, where the outcome is yet unknown.
- Dramatic – the size of the disaster, accident, earthquake, storm, robbery.
- Geographically near – the closer it is, the smaller it needs to be to affect me.
- Culturally relevant – I may feel connected to even a distant incident if I have something in common with it.
- Immediate – events rather than trends.
- Novel – the unusual or coincidental as they affect people.

On a different scale, sport can be all of these.

News has been called 'The Mirror of Society'. But mirrors reflect the whole picture, and news certainly does not do that. Radio news is highly selective, by definition it is to do with the unusual and abnormal but the basis of news selection must not be whether a story arouses curiosity or is spectacular, but whether it is significant and relevant. This certainly does not mean adopting a loftily worthy approach – dullness is the enemy of interest – it is to find the right point of human contact in a story. This may mean translating an obscure but important event into the listener's own understanding. A sharp change in the money markets will be readily understood by the specialist, but radio news must enable its significance to be appreciated by the man in the street. The job of news is not to shock but to inform. A broadcasting service will be judged as much by what it omits as what it includes.

Investigative reporting

The investigation of private conduct and organisational practice – and malpractice – is an important part of media activity. Newspapers have long regarded themselves as watchdogs, keeping an eye particularly on those in positions of public trust. The role of *The Washington Post* in the Watergate exposé is a well-documented example. Radio too recognises that it is not enough to wait for every news story to break of its own accord – some, of genuine public concern, are protected from exposure simply because of the vested interests which work to ensure that the truth never gets out. It is therefore sometimes necessary to allocate newsroom effort to the process of research enquiry into a situation which is not yet established fact. The story may never materialise because insufficient fact comes to light. This will involve the station in some loss through unproductive effort, but it is nothing like the loss which will be suffered if the newsroom proceeds to broadcast a story of accusations which turn out to be false.

Government departments or commercial businesses involved in underhand dealings; public officials or others with power engaged in questionable financial practice; the rich and famous called to account for their sexual immorality; these are the most common areas of investigation. But who is to say what is underhand, questionable, or immoral? While it may be possible to remain impartial in the reporting of news fact, the exposé inevitably carries with it an assessment of a situation against certain norms of behaviour. Such values are seldom purely objective. An investigation into the payment of a bribe in order to secure a contract may provoke a public scandal in one part of the world while in another it is simply the way in which business is conducted. In other words, investigation requires a judgement of some malpractice – of right and wrong. The reporter must therefore be correct on two counts – that the facts as reported stand up to later scrutiny, and that his or her own judgement as to the morality of the issue is subsequently endorsed by the listener, that is by Society.

To enable the reporter's own values to remain largely outside the investigation, the most fruitful approach is often to use the stated values of the organisation or person being investigated as the basis of the judgement made. Thus, a body claiming to have been democratically elected but which subsequently was shown to have manipulated the polls lays itself open to criticism by its own standards. The same is true of governments which, while happy to be signatories of agreements on the treatment of prisoners, also allow their armed forces to practise beatings and torture; or business firms which promise refunds in the event of customer complaint but which somehow always find a loophole to evade this particular responsibility. The radio station may have to represent the listener in cases of personal unfairness, or pursue the greater interests of society in the face of public corruption. But the broadcaster must be right. This takes patience, hard, wearying research, and the ability to distinguish relevant fact from a smokescreen of detail.

Occasionally, outside pressure will be brought to bring the enquiry to a halt. This may be the signal that someone is getting uncomfortable and that the effort is beginning to bear results. It is surprising how often malpractice breeds dissatisfaction. Once the fact of an investigation becomes known a person with a grudge is likely to provide anonymous information. Such 'leaks' and tip-offs of course need to be checked and treated with the utmost caution. A story that is told too soon will fall apart as surely as one that is wrong. Further, a station must resist the temptation to get so involved with a story that it falls prey to the same malpractice – although perhaps on a much smaller scale – as it is attempting to expose. So does it pay for information? Does it go in for surreptitious recording, of phone calls, for instance? Does it jeopardise its own integrity by giving false information or staging events in the hope of laying a trap for others? Investigative reporters should not work alone, but in twos or threes – to argue through the methods, develop theories, and assess results. They will be wise to stay in close contact with their management – whose backing and continuing financial support is crucial.

When it works, the effects are immediate and considerable. The reputation gained by the programme and station are incalculable. A 'scoop' puts competitors nowhere. The public at large wants wrongs put right. People respect a moral order, especially for others, and in the end prefer justice to expediency.

Campaigning journalism

Programmes, and their station or network, cease to be objective or impartial when they wholeheartedly promote a particular course of action. The extent to which any such bias will threaten the credibility of the station as a whole depends on the proportion of the audience which will already agree with the action proposed. Thus a local station which advocates a by-pass for its town, or wants to raise money for handicapped children, is unlikely to create opposition among its listeners. Even if the newsroom originates the campaign, the standing of the normal bulletin material will probably remain unaffected. If, however, the station is advocating action on a more contentious issue – non-smoking in all public places, the introduction of random breath checks as a deterrent to drunken drivers, or mandatory blood sampling to detect carriers of the AIDS virus – then the station must expect opposition, some of which will criticise any story on the subject which the station carries.

In general, campaigning is best kept away from the newsroom. The news editor should be allowed to pursue the professional reporting of daily factual truth without being involved in considerations of what other people – governments, councils, advertisers or radio managements – want reported, or unreported. This at least minimises the danger of one programme's editorial policy jeopardising news credibility. Voices associated with news almost always run some risk when they appear in another broadcasting context. Journalists who lend weight to a particular view, however worthy, easily damage their reputation as dispassionate observers.

A producer wanting to promote a cause must obviously seek the backing of his management and be aware of the possible effect of any campaign on other programmes, especially news. Partiality of view in itself may become counter-productive to the very issue it is supposed to promote. Political causes are the most extreme, and produce the greatest cynicism . . . 'but then you would say that, wouldn't you'.

The news reporting function

The reporter out on the street and the sub-editor at his desk are the people who make the decisions about news. Their concern is accuracy, intelligibility, legality, impartiality and good taste.

Before looking at the key principles it is important to say something about one of the more difficult aspects of the job.

Most of the work is relatively straightforward; some of it is routine. Chronicling events and the reasons for them requires much rewriting of other people's copy received by a multitude of means. It entails hours spent on the phone checking sources, and days out on location recording interviews and filing reports. It is during these times away from the newsroom, when the reporter is on his own, that he is required to have a sense of self-sufficiency – an apparent self-confidence, not always felt – to tackle the unknown and sometimes dangerous situation.

Civil disturbance or war reporting

Tragedy should be reported in a sombre manner – the broadcaster always remaining sensitive to how the listener will react. Reporting on a riot or commentating from a battle zone it is the reporter's task to report, not to get involved. It is sensible therefore always to get local advice on conditions and as far as possible to stay outside a disturbance, rather than try to work from inside the melee. It is then possible to see and assess what is happening as the situation develops. Under these conditions the reporter should remain as inconspicuous as possible and not add to or inflame the situation by his or her presence.

The primary ambience in a crisis is likely to be one of confusion. Asking for an official view tends to produce either optimistic hopes or worst fears, so any comment of this kind should be accurately attributed, or at least referred to as 'unconfirmed'. Analysis and interpretation of an event takes two forms – the pressures and causes which led up to it, and the implications and consequences likely to stem from it. Unless the reporter is very familiar with the situation it is best to leave reasons and forecasts to a later stage, and probably to others. On the spot, there is no room for speculation: the story should be told simply on the basis of what the reporter sees and hears.

In actual hostilities an accredited war reporter will be required to wear a flak jacket or other protective clothing – the military do not like to be held responsible unnecessarily for their own civilian casualties. It will be necessary to liaise closely with the officer in charge and to accept limits sometimes on what can be said. Facts may have to be withheld in the interests of a specific operation, for example the size and intent of troop movements, or the names and identities of people involved in a kidnapping or police siege. This is for fairly obvious tactical reasons and it is generally permissible to say that reporting restrictions are in force. One of the most memorable reports to come out of the Falkland Islands conflict arose from just such a situation. Brian Hanrahan reporting from the deck of the British aircraft carrier 'Hermes':

> At dawn our Sea Harriers took off, each carrying three one-thousand pound bombs. They wheeled in the sky before heading for the islands, at that stage just ninety miles away. Some of the planes went to create more havoc at Stanley, the others to a small airstrip called Goose Green near Darwin, 120 miles to the west. There they found and bombed a number of grounded aircraft mixed in with decoys. At Stanley the planes went in low, in waves just seconds apart. They glimpsed the bomb craters left by the Vulcan and left behind them more fire and destruction. The pilots said there'd been smoke and dust everywhere punctuated by the flash of explosions. They faced a barrage of return fire, heavy but apparently ineffective. I'm not allowed to say how many planes joined the raid, but I counted them all out and I counted them all back. Their pilots were unhurt, cheerful and jubilant, giving thumbs-up signs. One plane had a single bullet-hole through the tail – it's already been repaired.
> (Courtesy BBC News.)

Expressed in a cool unexcited tone, it is worth noting the shortness of the sentences and ordinariness of the words used. It is not necessary to use extravagant language to be memorable. See also the section on live commentary in Chapter 17.

Working in conditions of physical danger, a basic knowledge of first aid is valuable. Several organisations equip their staff reporting from areas of potential risk with a medical pack containing essentials such as sterile syringes, needles and intravenous fluid. Psychological as well as bodily safety remains important and reporters faced with violence, and the mutilated dead and dying – whether it be the result of a distant war, or a domestic train crash – can suffer trauma for some time afterwards as a result of their experiences. The sometimes harrowing effects of news work should not be underestimated and the opportunity always provided for suitable counselling.

Accuracy

A reporter's first duty is to get the facts right. Names, initials, titles, times, places, financial figures, percentages, the sequence of events – all must be accurate. Nothing should be broadcast without the facts being double checked, not by hearsay or suggestion but by thorough reliability. 'Return to the source' is a useful maxim. If it is not possible to check the fact itself, at least attribute the source declaring it to be a fact. Under pressure from a tight deadline, it is tempting to allow the shortage of time to serve as an excuse for lack of verification. But such is the way of the slipshod to their ultimate discredit. Even in a competitive situation, the listener's right to be correctly informed stands above the broadcaster's desire to be first. The radio medium, after all, offers sufficient flexibility to allow opportunity for continuing intermittent follow-up. Indeed, it is ideally suited to the running story.

Sometimes accuracy by itself is not enough. With statistics the story may be not in their telling but in their interpretation. For instance, according to the traffic accident figures the safest age group of motorbike users is the 'over 80s' – not one was injured last year! So a story concerning a 20 per cent increase in the radioactivity level of cows' milk over two years may be perfectly true, but is it significant? How has it varied at other times? Was the level two years ago unusually low? Were the measurements on precisely the same basis? And so on. Statistical claims need care.

Accuracy is required too in the sounds which accompany a report. The reporter working in radio knows how atmosphere is conveyed by 'actuality' – the noise of a building site, the shouts of a demonstration. It is important in achieving impact and establishing credibility to use these sounds, but not to make them 'bigger' than they really are. How fair is it then to add atmospheric music to an interview recorded in an otherwise silent café? It may be typical of the café's music (and it is useful in covering the edits) – but is it honest and real? Is it right to add small-arms fire to a report from a battle area? Typically it is there, it was just that the guns happened to be silent when I was recording. In other words does the piece have to *be* reality, or to convey reality? The moment you edit you destroy real-time accuracy. It is a question of motive. The accurate reporter, as opposed to a merely sensationalist one, will need a great deal of judgement if he is to excite and interest, but not mislead.

The basic structure for the news interviewer is first to get the facts, then to establish the reasons or cause lying behind them, and finally to arrive at their implication and likely resultant action. These three areas are simply past, present, and future – 'What's happened? Why do you think this is? What will you do next?' At another level, a news story is to do with the personal motives for decision and action, and it is these which have to be exposed, and if need be challenged with accurate facts or authoritative opinion quoted from elsewhere.

Intelligibility

Conveying immediate meaning with clarity and brevity is a task which requires refinement of thought and a facility with words. The first requisite is to understand the story so that it can be told without recourse to scientific, commercial, legal, governmental or social gobbledegook which so often surrounds the official giving of information. A reporter determined to show that he is at home with such technicalities through their frequent application has little use as a communicator. He must be the translator of jargon not its disseminator.

In recognising where to start he must have an insight into how much the listener already knows and how ideas are expressed in everyday speech. In being understood, the reporter's second requirement is therefore a knowledge of the audience – it is unwise to deal only with colleagues and professional sources for he will find himself subconsciously broadcasting only to them. If the audience is distant rather than local he must from time to time travel among them, or at the very least set up whatever means of feedback is possible.

The third element in telling a story is that it should be logically expressed. This means that it should be chronological and sequential – cause comes before effect:

Not: A reduction in the permitted level of cigarette advertising is recommended in a Department of Health report out today.

But: In a report out today the Department of Health recommends a reduction in the permitted level of cigarette advertising.

The key to intelligibility therefore is in the reporter's own understanding of the story, of the listener, and of the language of communication.

Putting these three aspects together the news writer's job is to tell the story as he understands it, putting it in a logical sequence, answering for the listener questions such as:

'What has happened?'
'When and where did it happen?'
'Who was involved?'
'How did it happen and why?'

The first technique is to ensure that of these six basic questions, at least three are answered in the first sentence:

1   The Chancellor of the Exchequer told Parliament this afternoon that he would be raising income tax by an average of 4 per cent in the autumn.

(who, where, when, what, when)

2   Eight people were killed and over sixty seriously injured when two trains collided just outside Amritsar in northern India in the early hours of this morning.

(what, where, when)

The second and subsequent sentences should continue to answer these questions:

1   He said it would be applied only to the upper tax bands and would not affect the basic rate. In answer to an Opposition question he said this was necessary to reduce the Government deficit.

(how, who, why)

2   The crowded overnight express from Delhi was derailed and overturned by a local freight train as it left a siding near Amritsar station. Railway workers and police are still taking the injured from the wreckage and it is feared that the death toll may rise.

(how, what)

A fault commonly heard on the air which we looked at in Chapter 5 is that of the misplaced participle.

'The Prime Minister will have to defend the agreement he signed, in the House of Commons.'

Without the comma this sounds as though we are talking about an agreement he signed in the House of Commons. But no, the story means

'The Prime Minister will have to defend in the House of Commons the agreement he signed.'

This error is frequently made in references to time.

'He said there was no case to answer last July.'

When what was actually meant was

'He said last July there was no case to answer.'

And another, following a report of child abuse by a nanny.

'There was a demand to register all nannies with Local Councils.'

But what, you might ask, about nannies who are not with Local Councils? It would be better as

'There was a demand for all nannies to be registered with Local Councils.'

Radio requires intelligibility to be immediate and unambiguous.

Legality

To stay within the law demands a knowledge of the legal process and of the constraints which the law imposes on anyone, individual or radio station, to say what they like. In Britain no one, for example, is allowed to pre-judge a case, to interfere with a trial, influence a jury or anticipate the findings. Thus there are considerable restrictions on what can be reported while a matter is *sub-judice.* To exceed the defined limits is to run the very severe risk of being held in contempt of court – an offence which is viewed with the utmost seriousness since it may threaten the law's own credibility.

Under present British law the outline of what is permissible in reporting a crime falls under four distinct stages:

1   *Before an arrest is made* it is permissible to give the *facts* of the crime but the description of a death as 'murder' should only be used if the police have made a statement to that effect. Witnesses to the crime may be interviewed but they must not attempt to describe the identity of anyone they saw or speculate on the motive.
2   After an arrest is made, or if a warrant for arrest is issued, the case is said to be 'active'. This continues *while the trial is in progress* and it is not permissible to report on committal proceedings in a magistrate's court, other than by giving the names and addresses of the parties involved, the names of counsel and solicitors, the offence with which the defendant is charged, and the decision of the court. The reporting of subsequent proceedings in the higher court is permitted but no comment is allowed. The matter ceases to be 'active' on sentence or acquittal.
3   Responsible comment is permissible *after the conviction and sentence is announced*, so long as the judge is not criticised for the severity or otherwise of the sentence, and there is no allegation of bias or prejudice.
4   *If an appeal is lodged,* the matter again becomes *sub-judice.* No comment or speculation is allowed and only factual court reports should be broadcast.

Complications can arise if the police are too enthusiastic in saying that 'they have caught the person responsible'. This is for the court to decide and broadcasters should not collude with police in pre-judging a case. There are special rules which apply to the reporting of the juvenile and matrimonial courts. The key question throughout is whether what is broadcast is likely to help or hinder the police in their investigation or undermine the authority of the judicial process.

Such matters are the stock in trade of the journalist, and producers unacquainted with the courts are advised to proceed carefully and to seek expert advice.

The second great area of the law of which all programme makers must be aware is that of libel. The broadcaster enjoys no special rights over the individual and is not entitled to say anything which would 'expose a person to hatred, ridicule or contempt, cause him to be shunned or avoided, or tend to injure him in his office, profession or trade'. To be upheld, a libel can only be committed against a clearly identifiable individual or group. In civil law, it is not possible to defame the dead. The most damaging accusation that can be brought against a broadcaster standing under the threat of a libel action is that he acted out of malice. This is not an unknown hazard for the investigative journalist working, for example, on a story about the possibility of corruption or dubious practice involving well-known public figures. The broadcaster's complete real defence against a charge of libel is that what he said was true, and that this can be proved to the satisfaction of a jury. Again, we have the absolute necessity of checking the facts and using words with a precision which precludes a possibly deliberate misconstruction.

A second defence is that of 'fair comment'. This means that the views expressed were honestly held and made in good faith without malice. This attaches particularly to book reviews, or the critical appraisal of plays and films, but may also apply to comments made about politicians or other public figures. Such a defence also has to show that the remarks are based on demonstrable facts not misinformation.

To repeat a libellous statement made by someone else is no defence unless that person enjoys 'absolute privilege', as in a court of law or in parliament. Even so, reports of such proceedings have to be fair as well as accurate and if the statement made turns out to be wrong and an apology or correction is issued, this too is bound to be reported. A defence of 'qualified privilege' is available to reports of other public proceedings such as local authority council meetings, official tribunals, company annual general meetings open to the public, and other meetings to do with matters of public concern. The same defence may be used in relation to a fair and accurate report of a public notice or statement issued officially by the police, a government department or local authority. Where no 'privilege' exists, the broadcaster is as guilty as the actual perpetrator of the libel. Producers and presenters of the phone-in should be constantly on their guard for the caller who complains of shoddy workmanship, professional incompetence, or worse, on the part of an identifiable individual. An immediate reference by the presenter to the fact that 'well, that's only your view' may be regarded as a mitigation of the offence, but the broadcaster can nevertheless be held to have published the libel.

The law also impinges directly on the broadcaster in matters concerning 'official' secrets, elections, consumer programmes, sex discrimination, race relations, gaming and lotteries, reporting from foreign courts, and copyright.

The individual producer should remain aware of the major legal pitfalls and must have a reliable source of legal advice. Without it he is likely, sooner or later, to need the services of a good defence lawyer.

Impartiality and fairness

The reporter does not select 'victims' and hound them – he does not ignore those whose views he dislikes, pursue vendettas, nor have favourites. He or she does not promote the policies of sectarian interests and resists the persuasions of those seeking free publicity. He is fair. Having no editorial opinion of his own, he seeks to tell the news without making moral judgements about it. He is the servant of his listener. Broadcasting is a general dissemination and no view is likely to be universally accepted. 'Good news' of lower trade tariffs for importers is bad news for home manufacturers struggling against competition. 'Good news' of another sunny day is bad news for farmers anxiously waiting for rain. The key is a careful watch on the adjectives, both in value and in size. Superlatives may have impact but are they fair? News may report an industrial dispute but what right does the reporter have to describe it as 'a *serious* industrial dispute'? On what grounds may he refer to a company's '*poor* record', or a medical research team's '*dramatic* breakthrough'? Words such as 'major', 'crucial', and 'special' are too often used simply to convince people that the news is important. Much better to leave the qualifying adjectives to the actual participants and for the newswriter to let the facts speak for themselves.

Reporters are occasionally concerned that they may not be able to be totally objective since they have received certain inbuilt values from their upbringing and education. While it may be true that broadcasting has more of its fair share of people from middle-class families and with a college education, any imbalance or restriction which results is the problem not of the reporter but the editor in chief. The reporter need not be unduly concerned with his own unconscious motivations of background and experience, except to recognise that others may not share them. However, he must be aware of, and subdue, any conscious desire he may have to persuade others to think the same way. It is sensible to ensure that any significantly large ethnic group in the community is represented in the broadcasting staff.

Unlike the junior newspaper journalist where every last adjective and comma can be checked before publication, the broadcast reporter is frequently in front of the microphone on his own. To help guard against the temptation to insert his own views, reporters should not be recruited straight from school, but have as wide and varied a background as possible and preferably bring to the job some experience of work outside broadcasting.

Good taste

As with all broadcasting, news programmes have a responsibility to abide by the generally accepted standards of what listeners would regard as 'good taste'. There are two areas which can create special problems – giving offence and causing distress.

In avoiding needless offence there must first be a professional care in the choice of words. People are particularly sensitive, and rightly so, about descriptions of themselves. The word 'immigrant' means someone who entered a country from elsewhere, yet it tends to be applied quite incorrectly to people whose parents or even grandparents were immigrants. Human labels pertaining to race, religion or political affiliations must be used with especial care and never as a social shorthand to convey anything other than their literal meaning. Examples are 'black', 'coloured', 'muslim', 'guerrilla', 'southern', 'jewish', 'communist', etc. – used loosely as adjectives they tend to be more dangerous than as specific nouns.

The matter of giving offence must be considered in the reporting of sexual and other crimes. News is not to be suppressed on moral or social grounds but the desire to shock must be subordinate to the need to inform. The journalist must find a form of words which when spoken will provide the facts without causing embarrassment, for example in homes where children are listening. With print, parents may divert their children from the unsavoury and squalid, in radio an immediate general care must be exercised at the studio end. A useful guideline is for the broadcaster to consider how he would actually express the news to someone in his local supermarket, with other people gathered round.

More difficult is the assessment of what is good taste in the broadcasting of 'live' or recorded actuality. Reporting an angry demonstration or industrial dispute when tempers are frayed is likely to result in the broadcasting of 'bad' language. What should be permitted? Should it be edited out of the recording? To what extent should it be deliberately used to indicate the strength of feeling aroused? There are no set answers, the context of the event and the situation of the listener are both pertinent to what is acceptable. However, in using such material as news, the broadcaster must ensure that his motive is really to inform and not simply to sensationalise. It may be 'good copy', but does it genuinely help the listener in his understanding of the subject? If so it may be valid but the listener retains the right to react as he or she feels appropriate to the broadcaster's decision.

News of an accident can cause undue distress. It is necessary only to mention the words 'air crash' to cause immediate anxiety among the friends and relatives of anyone who boarded a plane in the previous 24 hours. The broadcaster's responsibility is to contain the alarm to the smallest possible group by identifying the time and location of the accident, the airline, flight number, departure point and destination of the aircraft concerned. The item will go on to give details of the damage and the possibility of survivors, but by then the great majority of air travellers will be outside the scope of the story. In the case of accidents involving casualties, for example a bus crash, it is helpful for listeners to know to which hospital the injured have been taken or to have a telephone number where they can obtain further information. The names of those killed or injured should not normally be broadcast until it is known that the next-of-kin have been informed.

A small but not unimportant point in bulletin compilation is the need to watch for the unintended and possibly unfortunate association of individual items. It could appear altogether too callous to follow a murder item with a report on 'a new deal for butchers'. Common sense and an awareness of the listener's sensitivities will normally meet the requirements of good taste but it is precisely in a multi-source and time-constrained process, which news represents, that the niceties tend to be overlooked.

A summary

To sum up news principles. Good journalism is based on an oft-quoted set of values – it must be accurate and truthful, it stems from observation and enquiry, and it must do more than react to events, in that attempting to be impartial and objective it must actively seek out and test views. It has to make sense of events for readers, listeners and viewers, resisting the pressures of politicians, advertisers and others who may wish to cast the world in a light favourable to their own interests or cause. Any society founded on democratic freedom of choice requires a free flow of honest news. It is totally pointless to run a broadcast news service unless it is trusted and believed.

The newsroom operation

Almost every broadcasting station ultimately stands or falls by the quality of its news and information service. Its ability to respond quickly and to report accurately the events of the day extends beyond just news bulletins. The newsroom is likely to represent the greatest area of 'input' to a station and as such it is the one source capable of contributing to the whole of the output. Unlike a newspaper which directs its energies towards one or two specific deadlines, a radio newsroom is involved in a continuous process. The main sources of news coverage can be listed as:

1 *Professional:* staff reporters and specialist correspondents, e.g. crime, local government; freelances and 'stringers'; computer, fax and wire services; news agencies; syndicating sources including other broadcasting stations; newspapers.
2 *Official:* government sources both national and local; emergency services such as police, fire and hospitals; military and service organisations; public transport authorities.
3 *Commercial:* business and commercial PR departments; entertainment interests.
4 *Public:* information from listeners, taxi drivers, etc.; voluntary organisations, societies and clubs.

There is a danger that in basing its news too much on press releases and handouts a station is too easily manipulated by government and business interests. The output begins to sound like the voice of the Establishment. An editor will become wary of material arriving by hand from an official source just before a major deadline. Of course, a handout provides 'good', one-sided information – that is its purpose. It needs to be evaluated and cross-examined and questions asked about implications as well as immediate effects. A newsroom must be more than just a processor of other people's stories. The same is true of lifting items from newspapers – always look for new angles, and follow up if a story has appeared elsewhere, develop it, don't run it as it is, take it further.

The heart of the newsroom is its diary. This may be in book form or held as computer data. As much information as possible is gleaned in advance so that the known and likely stories can be covered with the resources available. The first editorial meeting of each day will review that day's prospects and decide the priorities. Reporters will be allocated to the monthly trade figures, the opening of a new airport or trunk road, the controversial public meeting, the government or industry press conference, the arrival of a visiting head of state, or the publication of an important report. The news editor's job is to integrate the work of the local reporters with the flow of news coming in from the other sources available, balancing the need for international, national or local news. But the news editor must also consider how he would deal with the unforeseen – an explosion at the chemical works, a surprise announcement by a leading politician, a murder in the street. A newsroom, however, cannot wait for things to happen, it must pursue its own lines of enquiry, to investigate issues as well as report events.

Local newsrooms are sometimes tempted to select bulletin items in terms of geographical coverage – going for 20 stories representative of the area, rather than 10 items of more universal interest. This should be resolved in the form of a stated policy – that is, the extent to which a newsroom regards itself as serving several minorities as opposed to one audience of collective interest. The first is true of a newspaper where each reader selects his own item for consumption, the second is more appropriate to radio where the selection is done for him. Given considerably less 'space' the fewer stories have to be of interest to everyone.

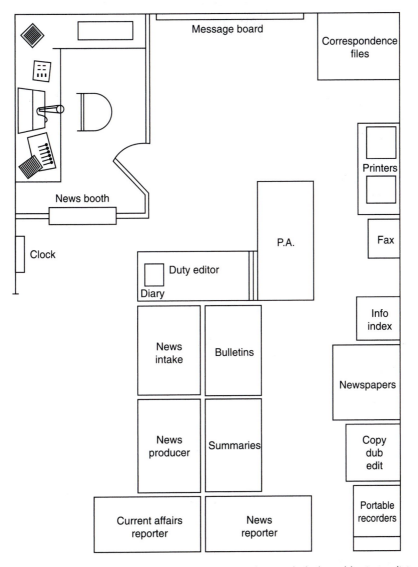

Figure 7.1 Features of a small newsroom. A terminal on each desk enables journalists to write and edit copy, gives access to all available sources, and provides for audio editing. Finished material is held on a central transmission store for subsequent playout. The news booth containing a digital audio workstation is used for making packages and telephone/ISDN interviews, as well as newsreading on air. Dubbing and editing facilities for cassettes, DAT, $^1/_4$-inch tape, and minidisc are provided as required

The competent newsroom has to be organised in its copy flow. There should be one place which receives the input of letters, press releases, printer and fax material, and other written data. A reporter making 'routine' calls to the emergency services or other regular contacts collects the verbal information so that, after consulting the diary, the news editor can decide which stories to cover. He may call a meeting in person or by phone to discuss the likely prospects, especially with specialist correspondents. One person will be allocated to write, or at least edit, the actual bulletins – a task not to be regarded as a committee job. If possible other writers are put on the shorter bulletins and summaries. Working from the same material, a two-minute summary needs a quite different approach. Omitting the last three sentences of each story in a five-minute bulletin will not do!

Reporters and freelances are allocated to the stories selected, each one is briefed on the implications and possible 'angles' of approach, together with suggestions for its treatment, and given a deadline for completion. Elsewhere in the room or nearby, material is edited, cue material written, recordings are made of interviews over the phone, and previous copy 'subbed', updated or otherwise refreshed for further use. The mechanical detail will depend on the degree of sophistication enjoyed by the individual newsroom – the availability of a radio car or other mobile direct inject equipment, OB and other 'remote' facilities, electronic data processes, off-air or closed circuit television, updated stories permanently available via computer screens in the studios, intercom to other parts of the building, etc.

In addition to all this the fully bi-media newsroom will not only be looking after the television side of things but may also have journalists to feed a variety of teletext, subtitle, and website services.

A newsroom also requires systems for the speedy retrieval of information from a central index. This may take the form of a physical file containing the names, addresses and telephone numbers of useful contacts, plus newspaper cuttings, scripts or other items relating to running stories or future events. These can be arranged either alphabetically or in date order but in such a way that everyone has access and understands the system. However, the computerised newsroom has the great advantage of being able to link together the diary information with 'today's' prospects, the current news programme running orders and bulletin scripts. Everyone has the same information updated at the same instant. The presenter in the studio has the latest news material constantly available, reading it from the screen. For the short-term retention of local information, the urgent telephone number or message to a colleague on another shift, a chalk board or similar device is simple and effective.

The important principle is that everyone should know exactly what they have to do to what timescale, and to whom they should turn in difficulty. The news editor or director in overall charge must be in possession of all the information necessary for him or her to control the output, and must be clear on the extent to which the minute-by-minute operation is delegated to others. There is no place or time for confusion or conflict.

Style book

One of the editor's jobs is to maintain the collection of rules, guidelines, procedures and precedents which forms the basis of the newsroom's day-to-day policy. It is the result of the practical experience of a particular news operation and the wishes of an individual editor. The style book is not a static thing but is altered and added to as new situations arise. A large organisation will issue guidelines centrally which its local or affiliate stations can augment.

Sections on policy will clarify the law relating to defamation and contempt of court. It will define the procedure to be followed, for example, in the event of bomb warnings (whether or not they are hoax calls), private kidnapping, requests for a news black-out, the death of heads of state, observance of embargoes, national and local elections, the naming of sources, and so on.

The book sets out the station's Mission or Purpose Statement and the role of news within that. It will indicate the required format for bulletins, the sign-on and sign-off procedures, the headline style, correct pronunciation of known pitfalls, and the policy regarding corrections, apologies and the right or opportunity of reply. It will list on-station safety regulations, as well as provide advice on proper forms of address. Above all there will be countless corrections of previous errors of grammar and syntax – from the use of collective nouns to the use of the word 'newsflash'.

The newcomer to a newsroom can expect to be given the style book on arrival and told to 'learn it'.

Radio car, mobile phone

The larger station will have a car, reserved for newsroom use, which the editor will send out with a reporter to cover a particular story. It will be used to interview the VIP, cover a demonstration, and report on a train crash. The very small operation may simply have a high-power radio microphone capable of sending a signal back to the station over a few kilometres. The principles of using mobile facilities tend to be similar whatever their design, and the following forms a common basis for routine operation:

- Ensure a proper priority procedure for every vehicle. Who controls and sanctions its use? Who decides if a booking is to be overridden to cover a more important story? Do all potential users know these procedures?

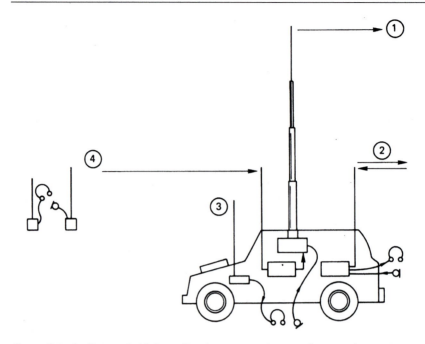

Figure 7.2 Radio car. 1. High-quality programme circuit to base studio. 2. Two-way communications link with base for cue and control. This enables the car to contribute to a programme not being broadcast 'live'. Where this link is used as the on-air programme circuit no other equipment is necessary. 3. Car radio receiver. 4. Optional low-power programme link enabling the reporter to work remotely from the vehicle if necessary, using a radio-mic and receiving off-air cue from an additional radio

- Remember that you are driving a highly distinctive vehicle. Whatever the hurry, be courteous, safe, and legal.
- Before leaving base, check that all necessary equipment is in the car, that the power supply batteries are in a good state of charge, and that someone is ready to listen out for you.
- Switch on the two-way communications receiver in the car so that the base station can call you. Tune the car radio to receive station output.
- On arrival at the destination, raise the telescopic mast where applicable, taking care to avoid obstructions and overhead power and telephone lines. Call the station on the communications link and send level on the programme circuit.
- Agree the cue to go ahead, duration of piece, and handback. Check that you are hearing station output on your headphones.
- After the broadcast, replace all equipment tidily and lower the mast before driving off (cars with telescopic masts should have safety interlocks to prevent movement with the mast extended). Leave the communications receiver on until back at base in case the station wishes to call you.
- Radio vehicles attract visitors; make sure vehicles are safely parked day and night.

Every car should be put 'on charge' when at base, and never garaged with the petrol tank less than half-full. It should carry area and local street maps, a reel of cable for remote operation, spare headphones, batteries and petrol can, extension leads, pad of paper, clipboard, pencil, cleaning cloth and torch.

A much cheaper radio car is a simple run-about vehicle equipped with a mobile phone. A digital phone, especially one adapted to take a broadcast quality microphone, can connect to the base studio using the public telephone system to provide programme inserts even from outside the station's transmission area. The reporter needs to hear cue programme in order to go ahead, and of course cannot be sent talkback while on air. But the advent of the mobile phone has greatly increased the ease and flexibility of location reporting.

Equipment in the field

Distance from the station of course creates no problems in the use of the Internet, either for sending text via e-mail or reports and interviews in digital audio form. The major broadcasting organisations typically use this method of communication with their network of news correspondents worldwide. A further development is ISDN – the Integrated Services Digital Network. At relatively low cost this system uses telephone circuits for high-quality voice transmission. The reporter plugs his microphone or recorder playback into an ISDN 'black box' (Codec) which encodes the signal in digital form. The system works well for speech and may be 'stretched' for music. Where phone lines are unreliable a digital recorder, laptop or notebook, and a mobile phone – or better still a satellite phone – is the only kit required to get fully edited and packaged interviews and reports immediately on air from anywhere in the world.

However, while the high-tech solution may seem attractive, the reporter working away from base soon learns to become self-reliant when it comes to his or her technical equipment. It is often the low-tech which saves the day – the Swiss army knife or the collection of small screwdrivers. Everything has a back-up. Experienced overseas reporters rely on such items as:

- Battery recording machines – mono cassette recorders are robust but a minidisc is good for quick editing. If one has a large loudspeaker it can be used for mixing acoustically with voice for recording on the second machine. Alternatively, a recorder with two mic/line inputs and internal mixing.
- Robust omni mics with built-in windshield.
- Small folding mic stand for use on a table.
- A lip mic to exclude untreated room acoustics or for use while travelling in a car.

- At least one long mic cable for putting a mic up-front for speeches or press conferences.
- Assorted plugs and leads, including crocodile clips for connecting to phone lines.
- Waterproof insulating tape, gaffer/sticky tape.
- Headphones or ear-piece for monitoring.
- International double adaptor telephone plug. This is useful to be able to turn one hotel telephone socket into two for simultaneous connection of a telephone and a recorder.
- A radio – FM/MW/LW/SW with extra wire for an aerial.
- Spare batteries, cassettes, discs, leads, etc.

Such kit is invariably carried in a tough foam-padded metal camera case – it is never left out of sight.

The news conference and press release

News conferences, company meetings, official statements and briefings of all kinds almost always set out to be favourable to those who hold them.

This is particularly so in the realm of politics. Ministerial aides, political advisers, press secretaries, public relations spokespersons, spin-doctors and the like. Their job is to create a positive appearance to all proposals and action – and to suppress the negative.

It is important therefore to listen carefully to what is being said – and what is not being said – and to ask key questions. Having been given facts or plans for action, the question is 'why?' It is necessary to quote accurately what is said – not difficult with a recorder – and to attribute the spokesperson or source. What is being said may or may not be totally true – what is true is that a named person is saying it.

Press releases, publicity handouts, notices and letters descend on the editor's desk in considerable quantity. Most end up on the spike, unused – although if it is not suitable for a bulletin, a well-organised newsroom will offer a story to another part of the output. From time to time editors are asked to define what should go into a press release and how it should be laid out. The following guidelines apply.

1   The editor is short of time and initially wants a summary of the story, not all the detail. Its purpose is to create interest and encourage further action.
2   At the top should be a headline which identifies the news story or event.
3   The main copy should be immediately intelligible, getting quickly to the heart of the matter and providing sufficient context to highlight the significance of the story – *why* it is of interest as well as *what* is

of interest, e.g. 'this is the first 18-year-old ever to have won the award'. The writing style should be more conversational than formal; avoid legal or technical jargon.

4 Double-check all facts: names (first name and surname), personal qualifications, titles, occupations, ages, addresses, dates, times, places, sums of money, percentages, etc.

5 End with a contact name, office of origin, address, telephone numbers (including a home number), date and time of issue.

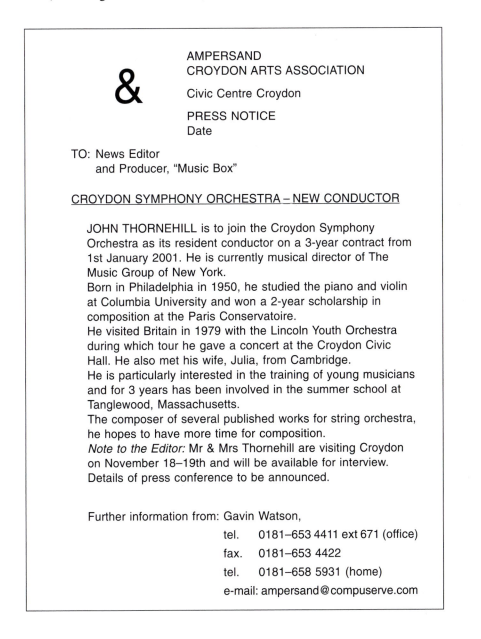

&

AMPERSAND
CROYDON ARTS ASSOCIATION

Civic Centre Croydon

PRESS NOTICE
Date

TO: News Editor
and Producer, "Music Box"

<u>CROYDON SYMPHONY ORCHESTRA – NEW CONDUCTOR</u>

JOHN THORNEHILL is to join the Croydon Symphony Orchestra as its resident conductor on a 3-year contract from 1st January 2001. He is currently musical director of The Music Group of New York.

Born in Philadelphia in 1950, he studied the piano and violin at Columbia University and won a 2-year scholarship in composition at the Paris Conservatoire.

He visited Britain in 1979 with the Lincoln Youth Orchestra during which tour he gave a concert at the Croydon Civic Hall. He also met his wife, Julia, from Cambridge.

He is particularly interested in the training of young musicians and for 3 years has been involved in the summer school at Tanglewood, Massachusetts.

The composer of several published works for string orchestra, he hopes to have more time for composition.

Note to the Editor: Mr & Mrs Thornehill are visiting Croydon on November 18–19th and will be available for interview. Details of press conference to be announced.

Further information from: Gavin Watson,

tel.	0181–653 4411 ext 671 (office)
fax.	0181–653 4422
tel.	0181–658 5931 (home)
e-mail: ampersand@compuserve.com

6 It should be typed, double-spaced with broad margins, on one side of a standard size such as A4, with a distinctive logo or heading – or on coloured paper to make it stand out. When sending copies to different addresses within a radio station, this should be made clear. An embargo should only be placed on a press release if the reason is obvious – for example, in the case of an advance copy of a speech, or where it is sensible to allow time to digest or analyse a complex issue before general publication. Radio is an immediate medium and editors have no inclination to wait simply in order to be simultaneous with their competitors.

8
Newsreading and presentation

Presentation is radio's packaging. It hardly matters how good a programme's content, how well written or how excellent its interviews; it comes to nothing if it is poorly presented. It is like taking a beautiful perfume and marketing it in a medicine bottle. Good presentation stems from an understanding of the medium and a basically caring attitude towards the listener. A broadcaster at the microphone ought consciously to care whether or not the listener can follow and understand what he is saying. By 'thinking outwards', away from himself, a newsreader or presenter is less prone to the destructive effects of studio nerves; or for that matter the complacency of overfamiliarity, and is therefore more likely to communicate meaning. Since he does not know the listener personally he adopts the relationship of an acquaintance rather than that of a friend. He is friendly, respectful, informative and helpful. He has something to offer the listener, but does not use this to gain advantage either by exercising a knowledgeable superiority or by assuming any special authority. The relationship is a horizontal one, he refers to his 'putting something across', not down or up. He does not make undue assumptions about his acquaintance, or presume on the relationship, but works at it always taking the trouble to make what he is saying interesting.

Of course, newsreading tends to be more formal than a music programme but there is room for a variety of approaches. Whatever the overall style of the station, governed by its basic attitude to the listener, it should be fairly consistent. While the sociologist may regard radio as a mass-medium, the person at the microphone sees it as an individual communication, he is talking to some*one*. Thinking of his listener as one person he says, 'If you're travelling south today, . . .' not 'Anyone travelling south . . .' The presenter does not shout. If he is half a metre from the microphone and his listener is a metre from the radio, the total distance between them is one and a half metres. What is required is not volume but an ordinary clarity. Too much projection causes the listener psychologically to 'back off' – it distances the relationship. Conversely, by dropping his voice the presenter adopts the confidential or intimate style more appropriate to the closeness of late night listening.

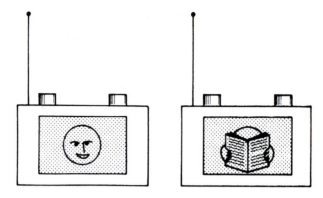

Figure 8.1   The script must not come between the broadcaster and the listener. The listener should feel that he is being spoken to – not read to. The script needs to be written for talking aloud and the vocal inflections and stresses kept as natural as the broadcaster's own speech

The rate of delivery depends on the style of the station and the material being broadcast. Inter-programme or continuity announcements should be at the presenter's own conversational speed, for example newsreading at 160–200 words per minute, but slower if short-wave transmission. Commentary should be at a rate to suit the action. If a reader is going too fast it may not help simply to ask him to slow down; this is likely to make him sound stilted and overcareful. What is required is that he leaves more pause between the sentences – that is when the understanding takes place. It is not so much the speed of the words which can confuse, but the lack of sufficient time to make sense of them.

The simplest way of getting the style, projection and speed right is to visualise the listener sitting in the studio a little way beyond the microphone. The presenter is not by himself reading, he is talking with the listener. This small exercise in imagination is the key to good presentation.

Newsreading

The first demand placed on the newsreader is that he understands what he is reading. He cannot be expected to communicate sensibly if he has not fully grasped the sense of it himself. With the reservations expressed later about 'syndicated' material, there is little place for the newsreader who picks up a bulletin with thirty seconds to go and hopes to read it 'word perfect'. He must be better than punctual, he must be early. Neither is technically faultless reading the same thing as communicating sense. A newsreader should be well informed and have an excellent background knowledge of current affairs so that he can cope when changes occur just before bulletins. He should take time to read it out loud beforehand – this gives him a chance to understand the content and be aware of pitfalls. There may be problems of pronunciation over a visiting Chinese trade mission or statement by an African foreign minister. There may be a phrase which is awkward to say, an ambiguous construction, or typing error. The pages should be verbally checked by the person who, in the listener's mind, is responsible for disseminating them. While a newsroom may like to give the impression that its material is the latest 'hot off the press' rush, it is seldom impossible for the reader to go through all the pages as a bulletin is being put together. Thorough preparation should be the rule, with reading at sight reserved for emergencies.

Of course in practice this is often a counsel of perfection. In a small station where the newsreader may be working single-handed, the news can arrive within seconds of its deadline. It has to be read at sight. This is not the best practice and runs a considerable risk of error. It places on the syndicating news service, and on the transmitting keyboard operator, a high responsibility for total accuracy. The reason for poor broadcasting of on-screen or 'rip 'n' read' material may lie with the station management for insufficient staffing, or with the news agency for less than professional standards. The fact of the matter is that in the event of a mistake on the air, from whatever cause, the listener blames the newsreader.

The person at the microphone therefore has the right to expect a certain level of service. This means a well-written and properly set out bulletin, accurately typed, arriving a few minutes before it is needed. He can then check to see if the lead story has changed and scan it quickly for any unfamiliar names. He picks out figures and dates to make sure he understands them. In the actual reading, his eyes are a little ahead of his speech, enabling him to take in *groups of words*, understanding them before passing on their sense to the listener.

The idea of syndicated news is excellent but it should not become the cause nor be made the excuse for poor microphone delivery.

In the studio the newsreader sits comfortably but not indulgently, feeling relaxed but not complacent, breathing normally and taking a couple of extra deep breaths before beginning.

Here are some other practicalities of script reading:

- Don't eat sweets or chocolate beforehand: sugar thickens the saliva.
- Always have a pen or pencil with you for marking alterations, corrections, emphasis, etc.
- If you wear them, make sure you have your glasses.
- Don't wear anything that could knock the table or rattle – bangles, cufflinks.
- Place a glass of water near at hand.
- Remove any staples or paper-clips from the script and separate the pages so that you can deal with each page individually.
- Make sure you have the whole script, check that the pages are in the right order, the right way up.
- Give yourself space, especially to put down the finished pages – don't bother to put them face down.

- Check the clock, cue light, headphones – for talkback and cue programme, and the mic-cut key if there is one.
- Check your voice level.
- Where timing is important, time the final minute of the script (180 words – perhaps 18 lines of typescript) and mark that place. You need to be at that point with a minute left to go and may have to drop items in order to achieve this.
- Once started, don't worry about your own performance, be concerned that you are really communicating to your imagined listener, 'just beyond the mic'.
- If reading from a computer screen connected to a local network system, check that a colleague on another terminal is not inputting to the news while you are broadcasting.

Pronunciation

A station should as far as possible be consistent over its use of a particular name. Problems arise when its output comprises several sources, e.g. syndicated material, a live audio news feed, a sustaining service. What must be avoided is one pronunciation in a nationally syndicated bulletin, followed a few minutes later by a different treatment in a locally read summary. The newsroom must listen to the whole of the station's news output, from whatever source, and advise the newsreader accordingly. Second, listeners are extremely sensitive to the incorrect pronunciation of names with which they are associated. The station which gets a local place name wrong loses credibility, one which mispronounces a personal name is regarded as either ignorant or rude. The difficulty is that listeners themselves may not agree on the correct form. Nevertheless a station should make strenuous efforts to ensure a consistent treatment of place names within its area. A phonetic pronunciation list based on 'educated local knowledge' should be adopted as a matter of policy and a new broadcaster joining the staff acquainted with it at the earliest possible time.

Alternatively it may be possible to store correctly spoken pronunciations in audio form on a computer hard disk. It is then an easy matter for a presenter to bring a name up on the screen and to hear it being said.

Vocal stressing

An important aspect of conveying meaning about which a script gives no clue is that of stress – the degree of emphasis laid on a word. Take the phrase:

'What do you want me to do about that?'

With the stress on the 'you', it is a very direct question. On the 'me', it is more personal to the questioner; on the 'do', it is a practical rather than a theoretical matter; on the 'that', it is different again. Its meaning changes with the emphasis. In reading news such subtleties can be crucial. For example, we may have in a story on Arab/Israeli affairs the following two statements:

> Mr Radim is visiting Washington where he is due to see the President this afternoon. Meanwhile the Israeli Foreign Minister is in Paris.

The name is fictional but the example real. Put the emphasis on the word 'Israeli', and Mr Radim is probably an Arab foreign minister. Put it on 'Foreign', and he becomes the Israeli Prime Minister. Try it out loud. Many sentences have a central 'pivot', or are counter-balanced about each other: 'While *this* is happening over *here*, *that* is taking place over *there*.' Many sentences contain a counter-balance of event, geography, person or time: 'Mr *Smith* said an election should take place *now*, *before* the issue came up. Mr *Jones* thought it should wait at least until *after* the matter had been debated.'

Listening to newsreaders it is possible to discern a widespread belief that there is a universal news style, where speed and urgency have priority over meaning, where the emphasis is either on every word or scattered in a random fashion, but always on the last word in every sentence. Does it stem from the journalist's need for clarity when dictating copy over the phone? The fact is that a single misplaced emphasis will cloud the meaning, possibly alter it. The only way of achieving correct stressing is by fully understanding the implications as well as the 'face value' of the material. This must be a conscious awareness during the preparatory read-through. As has been rightly observed, 'take care of the sense and the sounds will take care of themselves'.

Inflection

The monotonous reader either has no inflection in his voice at all or the rise and fall in pitch becomes regular and repetitive. It is the predictability of the vocal pattern which becomes boring. A too typical sentence 'shape' starts at a low pitch, quickly rises to the top and gradually descends, arriving at the bottom again by the final full stop. Placed end to end, such sentences quickly establish a rhythm which, if it does not mesmerise, will confuse because with their beginning and ending on the same 'note', the joins are scarcely perceptible. Meaning begins to evaporate as the structure disappears. Without sounding artificial or contrived, sentences normally start on a higher pitch than the one on which the previous one ended – a new paragraph certainly should. There can often be a natural rise and fall within a sentence, particularly if it contains more than one phrase. Meaningful stressing rather than random patterning will help.

A newsreader is well advised occasionally to record his work for his own analysis – is it too rhythmic, dull or aggressive? In the matter of inflection he should experiment off the air, putting a greater rise and fall into his voice than usual to see whether the result is more acceptable. Very often when he may feel he is really 'hamming it up', the playback sounds perfectly normal and only a shade more lively. Even experienced readers can become stale and fall into the traps of mechanical reading, and a little non-obsessive self-analysis and experimentation is very healthy.

Quotation marks

Reading quotes is a minor art on its own. It is easy to sound as though the comment is that of the newsreader, although the writing should avoid this construction. Some examples:

> While an early bulletin described his condition as 'comfortable', by this afternoon he was 'weaker'. (This should be rewritten to attribute both quotes.)
>
> The opposition leader described the statement as 'a complete fabrication designed to mislead'.
>
> He later argued that he had 'never seen' the witness.

To make someone else's words stand out as separate from the newsreader's own, there is a small pause and a change in voice pitch and speed for the quote.

Alterations

Last-minute handwritten changes to the typed page should be made with as much clarity as possible. Crossings out should be done in blocks rather than on each individual world. Lines and arrows indicating a different order of the material need to be bold enough to follow quickly and any new lines written clearly at the bottom of the page. To avoid confusion a 'unity of change' should be the aim. It is amazing how often a reader will find his way skilfully through a maze of alterations only to stumble when his concentration relaxes on the next perfectly clear page.

Corrections

But what happens when a mistake is made? Continue and ignore it or go back and correct it? When is an apology called for? It depends of course on the type of error. There is the verbal slip which it is quite unnecessary to do anything about, a misplaced emphasis, a wrong inflection, a word which comes out in an unintended way. The key question is 'could the listener have misconstrued my meaning?' If so, it must be put right. If there is a persistent error, or a refusal of a word to be pronounced at all, it is better to restart the whole sentence. Since 'I'm sorry I'll read that again' has become a cliché, something else might be preferred – 'I'm sorry, I'll repeat that', or 'Let me take that again'. It is whatever comes most naturally to the unflustered reader. To the broadcaster it can seem like the end of the world; it is not. Even if the listener has noticed it he simply wants a correction with as little fuss as possible.

Lists and numbers

The reading of a list can create a problem. A table of sports results, stock market shares, fatstock prices or a shipping forecast; these can sound very dull. Again, the first job for the reader is to understand his material, to take an interest in it so that he can communicate it. Second, the inexperienced reader must listen to others, not to copy them, but to pick up the points in their style which seem right for himself. There are particular inflections in reading this material which reinforce the information content. With football results, for example, the voice indicates the result as it gives the score. The same is true of racing results, which have a consistent format:

> 'Racing at Catterick – the three thirty.
> First, number 7, Phantom, 5 to 2 favourite,
> Second, number 9, Crystal Lad, 7 to 1
> Third, number 3, Handmaiden, 25 to 1
> Non-runner, number 1, Gold Digger.'

For obvious reasons care must be taken over pronunciation and prices. A highly backed horse might be quoted at '2 to 5'. This is generally given as '5 to 2 on'.

In passing, it is worth noting that sport has a good deal of its own jargon which looks the same on the printed page, for instance the figure '0'. The newsreader should know when this should be interpreted as 'nought', 'love', 'zero', 'oh', or 'nil' – the listener will certainly know what is correct. Unless it is automatic to the reader, it is well worth writing the appropriate word on the script. Whenever figures appear in a script, the reader should sort out the hundreds from the thousands and if necessary write the number on the page in words.

If it has been correctly written, a script consists of short unambiguous sentences or phrases, easily taken in by the eye and delivered vocally well within a single breath. The sense is contained not in the single words but in their grouping. To begin with we learn to read letter by letter, then word by word. The intelligent newsreader delivers his material phrase by phrase, taking in and giving out whole groups of words at a time, leaving little pauses between them to let their meaning sink in. The overall style is not one of 'reading' – it is much more akin to 'telling'.

In summary, the 'rules' of newsreading are:

1   Understand the content by preparation.
2   Visualise the listener by imagination.
3   Communicate the sense by telling.

Station style

Radio managers become paranoid over the matter of station style. They will regard any misdemeanour on-air as a personal affront, especially if they instituted the rule which should have been observed. It's nevertheless true that a consistent station sound aids identification. It calls for some discipline, particularly in relation to the frequently used phrases to do with time. Is it 3.25 or 25 past 3? Is it 3.40 or 20 minutes to 4? Is it 1540? Dates: is it 'May the eleventh', 'the eleventh of May', or 'May eleven'? Frequencies, 1107 kilohertz, 271 metres medium wave, can be given in several ways: 'on a frequency of eleven-oh-seven', or 'two-seven-one metres medium wave', or 'two-seventy one'. Temperatures: Celsius, centigrade or Fahrenheit? Or just '22 degrees'?

Some stations insist on a very strict form of identification, some prefer variety:

Radio Berkely
Berkely radio – two seven one
The county sound of Berkely
The sound of Berkely county
Berkely county radio etc. etc.

Idents can be by station name, frequency or wavelength, programme title, presenter name, or by some habitually used catchline:

XFM – where news comes first
XFM – the heartbeat of the county
XFM – with the world's favourite music

Find out the station policy and stick to it – even when sending in an audition tape, use the form you hear on-air.

A frequent rule of presentation is 'never say goodbye'. It's an invitation for the listener to respond and switch off. At the end of a programme the presenter hands over to someone else – you (the station) never give the impression of going away. Further, the presenter joins the listener not the other way round. 'It's good to be with you', is a subtle form of service, whereas 'Thank you for joining me' is more of an ego-trip for the presenter. The station should go to the bother of reaching out to its listeners, not expect them to come to it.

Continuity presentation

Presenting a sequence of programmes, giving them continuity, acting as the voice of the station, is very similar to being the host of a magazine programme responsible for linking different items. The job is to provide a continuous thread of interest even though there are contrasts of content and mood. The presenter makes the transition by picking up in the style of the programme that is finishing so that by the time he has done the back-announcements and given incidental information, station identification and time check, he is ready to introduce the next programme in perhaps a quite different manner. Naturally to judge the mood correctly he has to do some listening. It is no good coming into a studio with under a minute to go, hoping to find the right piece of paper so as to get into the next programme, without sounding detached from the whole proceedings. A station like this might as well be automated.

If there is time at programme breaks, trail an upcoming programme – not the next one since you are going to announce that in a moment. The most usual style is to trail the 'programme after next'. But do so in a compelling and attractive way so as to retain the interest of the listener – perhaps by using an intriguing clip from the programme (see p. 119). If the trail is for something further ahead, then make this clear – 'Now looking ahead to tomorrow night . . .'

Continuity presentation requires a sensitivity to the way a programme ends, to leave just the right pause, to continue with a smile in the voice or whatever is needed. Develop a precise sense of timing, the ability to talk rather than 'waffle', for exactly fifteen seconds, or a minute and a half. A good presenter knows it is not enough just to get the programmes on the air, his concern is the person at the end of the system.

Errors and emergencies

What do you do when the computer fails to respond, the machine does not start or having given an introduction there is silence when the fader is opened? First, no oaths or exclamations! The microphone may still be 'live' and this is the time when one problem can lead to another. Second, look hard to see that there has not been a simple operational error. Are all the signal lights showing correctly? Is there an equipment fault which can be put right quickly? Can a recording be transferred to another machine? If by taking action the programme can continue with only a slight pause, five seconds or so, then no further announcement is necessary. If it takes longer to put right, ten seconds or more, something should be said to keep the listener informed:

'I'm sorry about the delay, we seem to have lost that report for a moment . . .'

Then if action is assured it is possible to continue:

'We'll have it for you shortly . . .'

The presenter may assume personal or collective responsibility for the problem but what he must not do is to blame someone else:

'Sorry about that, the man through the glass window here pressed the wrong button!'

The same goes for the wrong item following a particular cue, or pages read in the wrong order. The professional does not become self-indulgent, saying how complicated the job is, he simply puts it right, with everybody else's help, in a natural manner and with the minimum of bother. His job is to expect the unexpected.

Sooner or later a more serious situation will occur which demands that the presenter 'fills' for a considerable time. Standby announcements of a public service type – an appeal for blood donors, safety in the home, code for drivers, procedure for contacting the police or hospital service. Also programme trails and other promotional material can be used. These 'fills' should always be available to cover the odd twenty seconds, and changed once they have been used.

Standby music is an essential part of the emergency procedure. Something for every occasion – a break in the relay of a church service, the loss of Saturday football, an under-run of a children's programme. To avoid confusion the music chosen should not be identical to anything it replaces, simply of a sympathetic mood. Once it is on there is a breathing space to attempt to get the problem sorted out. The principle is to return to the original programme as quickly as possible. Very occasionally it may be necessary to abandon a fault-prone programme and some stations keep a 'timeless talk' or fifteen-minute feature permanently on standby to cover such an eventuality. For longer-term schedule changes see p. 281.

Headphones

A vocal performer can sometimes become obsessed with the sound of his or her own voice. The warning signs include a tendency to listen to himself continuously on headphones. The purpose of headphone monitoring is essentially to provide talkback communication, or an outside source or cue programme feed. Only if it is unavoidable should both ears be covered, otherwise the presenter begins to live in a world of his own, out of touch with others in the studio. If he has a great deal of routine work, the same announcements, station identifications, time checks and introductions, it is easy not to try very hard to find appropriate variations. Like the newsreader he should occasionally listen to himself recorded off-air, checking a repetitive vocabulary, use of cliché or monotony of style.

Trails and promos

Part of a station's total presentation 'sound' is the way it sells itself. Promotional activity should not be left to chance but be carefully designed to accord with an overall sense of style. 'Selling' one's own programmes on the air is like marketing any other product, and this is developed in the next chapter, but remember that the appeal can only be directed to those people who are already listening. The task is therefore to describe a future programme as so interesting and attractive that the listener is bound to tune in again. The qualities which people enjoy and which will attract them to a particular programme are:

- Humour that appeals.
- Originality that is intriguing.
- An interest that is relevant.
- A cleverness which can be appreciated.
- Musical content
- Simplicity – a non-confusing message.
- A good sound quality.

If one or more of these attributes is presented in a style to which he can relate, the listener will almost certainly come back for more. The station is all the time attempting to develop a rapport with the listener, and the programme trailer is an opportunity to do just this. It is saying of a future programme, 'this is for *you*'.

Having obtained the listener's interest, a trail must provide some information on content – what the programme is trying to do, who is taking part, and what form (quiz, discussion, phone-in, etc.) it will take. All this must be in line with the same list of attractive qualities. But this is far from easy – to be humorous *and* original, to be clever as well as simple. The final stage is to be sure that the listener is left with clear transmission details, the day and time of the broadcast. The information is best repeated:

'. . . You can hear the show on this station tomorrow at six p.m. Just the thing for early evening – the 'Kate Greenhouse Saga', on 251 – six o'clock tomorrow.'

Trails are often wrapped around with music which reflects something of the style of the programme, or at least the style of the programme in which the trail is inserted. It should start and finish clearly, rather than on a fade; this is achieved by prefading the end music to time and editing it to the opening music so that the join is covered by speech.

At its simplest, a trail lasting 30 seconds looks like this:

```
MUSIC:        Bright, faded on musical phrase, held under speech.        5"

SPEECH:       Obtains interest.                                          10"
              Provides information on content.
              (Music edit at low level under speech)                    5"
              Gives transmission details.                               5"

MUSIC:        Fade up to end.                                           5"
```

There is little point in ordering a listener to switch on; the effect is better achieved by convincing him that he will be deprived if he does not. And of course if that is the station's promise then it must later be fulfilled. Trails should not be too mandatory, and above all they should be memorable.

9

Making commercials

The purpose of an advertisement is to sell things. It is not there simply to amuse people or to impress the chairman's wife; it is designed to move goods off shelves, cars out of showrooms, and customers eagerly towards services. The radio advertiser must use a good deal of skill in motivating his target audience to a specific action. The effective advertisement will:

- interest
- inform
- involve
- motivate
- direct.

Most commercials are made by advertising agencies in conjunction with specialist production houses. They arrive at the station on CD or cassette and need only transferring to the most suitable operational medium – generally a computer. However, producers may be called upon to make local commercials and the elements which he or she must consider are:

- The target audience – for whom is this message primarily intended?
- The product or service – what is the specific quality to be promoted?
- The writing – what content and style will be appropriate?
- The voice or voices – who will best reinforce the style?
- The background – is music or sound effects needed?

The producer must also be familiar with the station copy policy or code of practice governing advertising. Regulatory standards form the essential background to commercial production.

Copy policy

'I've bought the time. I can say what I like.' Unfortunately not. A client does not have free rein with the station's air but must comply with a set of rules, a copy of which should be readily available to all potential advertisers. In Britain the Broadcasting Act 1990 makes the issuing of such rules the duty of the Radio Authority, which is also charged with enforcing them.

The principle is that all advertising should be 'legal, decent, honest and truthful' and that nothing should 'offend against good taste or public feeling'.

The Radio Authority's Code of Advertising Standards and Practice, and Programme Sponsorship sets out the rules, including specific prohibitions on advertising which:

* could be confused with programming
* is on behalf of any political body
* shows partiality in matters of current political or industrial controversy
* unfairly attacks or discredits other products
* includes sounds likely to create a safety hazard for drivers
* makes use of product placement in programmes
* exploits the superstitious or plays on fear
* is on behalf of any body which practises or advocates illegal behaviour
* makes claims which give a misleading impression.

No advertising is permitted within coverage of a religious service, formal royal ceremony, or a schools' programme under 30 minutes. Programme presenters must not endorse a product in a presenter-read commercial. The Code lists those categories where central clearance of copy is required and includes detailed sections on financial advertising – investments, savings, insurance, etc. – alcohol, advertising for and by children, the advertising of medicines and treatments including contraceptives and pregnancy-testing services, charity advertising, appeals for disasters and advertising on behalf of religious bodies. It bans the advertising of cigarettes, guns and gun clubs, pornography, the occult, betting and gaming, escort agencies, products for the treatment of alcoholism, hypnosis and psychoanalysis. It also lists the current legislation relevant to broadcast advertising. The Code is essential reading for anyone involved in British commercial radio.

In the US radio advertising regulation is chiefly through the self-adopted code of the National Association of Broadcasters. This is similar to the Radio Authority rules quoted and has points to make about the use of words 'safe' and 'harmless' related to pharmaceuticals, the presentation of marriage, and the sensitivity necessary in the use of material relating to race, colour or ethnic derivation. The advertising of hard liquor (distilled spirits) is prohibited. In addition to the National Code, network and local stations have their own policies which conform to State laws. Some of these are very detailed, defining terms such as 'like new', 'biggest', 'factory fresh', 'no credit refused' and 'guarantee'.

Public Service Announcements made available to charitable and non-profit organisations without charge must conform to the same standards as the paid-for commercials and also require approval by the commercial department. Stations may define the standard of language acceptable – where payment is on wordage there is a temptation for advertisers to supply copy in 'telegraphese'. Copy policy is not a fixed and immutable thing. It goes out of date as standards and fashions vary. The commercial producer must therefore keep abreast of these regulations as he is aware of changes in the law itself.

The target audience

Only on the very smallest station will the producer be required to sell airtime. The qualities of persistence, persuasion and patience belong to the sales team who will negotiate the price, the number of spots, and the discount offered by the Rate Card. The placing of the spots will of course crucially affect the rate charged, and is something the producer also needs to know for two reasons. First, because the transmission time must be appropriate to the intended listener, and second, because it may affect the written copy. Clearly there is no point in broadcasting a message to children when they are in school, or to farmers when they are busy. Nothing is gained by exhorting people to 'buy something now' when the shops are shut. If you are selling holidays, who makes the purchasing decision for which type of vacation? What precise age buys what type of record? Which socio-economic group do you wish to attract for a particular drink?

To sell its airtime effectively, a commercial station needs to know about its audience. Through independent market research it has to show why it is better than its competitor at reaching a particular section of the public. What age groups does it attract? How much disposable income do they have, and what do they do with it? To what extent do they buy coffee, cars, furniture, holidays, magazines, mortgages or insurance? And of what kind – expensive upmarket or 'cut-price'? If you are the manufacturer of kitchen equipment you want to spend your advertising budget where it is most likely to pay off. So which is the best way of reaching newly married couples setting up home? A radio station with a well-researched audience profile is much more convincing than an amateur with a hunch.

The product or service 'premise'

In 30 seconds, it's not possible to say everything about anything. So identify one, or perhaps two, key features about the product which mark it out as especially attractive. Choose one of these – its usefulness, efficiency, simplicity, low cost, durability, availability, value for money, exclusivity, technical quality, newness, status, advanced design, excitement or beauty. There are other possibilities but a single memorable point about a product is far more effective than an attempt to describe the whole thing. In the case of food and drink the key phrases may be more subjective – easy to prepare, made in a moment, satisfying, long lasting, luscious, nutritious, and so on.

Now develop a short statement which connects the product's intention with a known and desirable effect. This becomes the 'critical premise' or consumer benefit, and often comprises a subject, an action verb and a result. Here are some examples:

> Clean breath helps personal relationships.
> Disinfecting your bathroom gives you a safer home.
> A slimming diet makes you more attractive.
> Serving rich coffee impresses your friends.
> Driving while drunk kills innocent people.
> Unsafe sex increases the risk of AIDS.
> Inoculation against disease can save your child's life.
> An energy breakfast promotes a successful day.

This is the writer's hypothesis and is the essential first building block for any radio spot. Before going further, test it out on other people – do they believe it? Does the action genuinely link the cause and the effect? Is the end result something that people in your target audience want? This brings us back to the point that effective advertising is grounded on thorough, relevant research? If the premise is shown to be true it is now necessary to connect the product firmly with it.

Whether the object of attention is a January sale, a fast food shop, a cosmetic cream or a life insurance policy, the client and producer/writer together must agree the primary feature to be sold.

It is important to consider the overall style or image to be projected. Is the impression required to be friendly, warm and domestic or is it unusual, lively and adventurous? If the idea is to convey reassurance, dependability and safety, this should be communicated in the writing, but also in the voicing and any music used. Every element must be consistent in combining to support the premise, and associate the product with it.

Writing copy

This is the heart of it, and it's worth remembering two things – (1) well-chosen, appropriate words cost no more than sloppy clichés; and (2) radio is a visual medium.

First impressions count – the 'primacy' and 'recency' effects referred to under 'News' are no less important here. The opening sentence should immediately identify the setting – the location, persona, key selling point or product. This can be greatly helped by well-chosen effects:

```
MALE VOICE CHOIR:    'All Through the Night' (faded, continues under)

FX:                  Clank of milk bottles

VOICE:               Down at the depot, before the crack of dawn,
                     You'll find the lads from Unigate preparing for the
                     morn.
                     They're loading up the turkeys, packing up the cheese,
                     the freshest cream, the choicest meats,
                     the tempting range of Christmas treats
                     that's surely bound to please.
                     (pause - music continues under)

                     Your Unigate milkman can offer you so much more than
                     milk this Christmas, and all at super-value prices. So
                     this Christmas don't forget to stop the milkman - he's
                     got a float full of Christmas goodies

                                                       - from Unigate.

FX:                  Owlhoot (music ends)              (55 secs)

[Courtesy County Sound Radio - station-produced commercial]
```

The American copywriter Robert Pritikin has pointed out the value of specifically writing for the eye as an aid to product recall. He wrote a now famous illustration of radio's ability to help the listener to visualise even something as intangible as a colour:

```
ANNCT:    The Fuller Paint Company invites you to stare with your ears at
          . . . yellow. Yellow is more than a colour. Yellow is a way of
          life. Ask any taxi driver about yellow. Or a banana salesman.
          Or a coward. They'll tell you about yellow. (Fx Phone rings)

          Oh, excuse me. Yello!! Yes, I'll take your order. Dandelions, a
          dozen; a pound of melted butter; lemon drops and a drop of
          lemon, and one canary that sings a yellow song. Anything else?
          Yello? Yello? Yello? Oh, disconnected. Well, she'll call back.

          If you want yellow that's yellow-yellow, remember to remember
          the Fuller Paint Company, a century of leadership in the
          chemistry of colour. For the Fuller colour center nearest you,
          check your phone directory. The yellow pages, of course.

[Robert C. Pritikin, writing in 'Monday Memo', Broadcasting
Magazine, 18 March 1974, p. 22]
```

Here the listener can put his own images to the ideas presented – and the key point about this paint, not its value or durability – its yellowness. Everything is geared to communicate its bright liveliness – even the shortness of the sentences. On reading the piece any good producer will be able to hear the appropriate voice for it and, if need be, the right music.

Creating something visual to produce a memorable image leading to product recall demands great imagination – especially for the more mundane. After all, as an on-station producer what would you write for a local windscreen repair service?

```
MAN'S VOICE:       You're not going to believe this.
                   (Music under – orchestral strings, urgent 'thriller'
                   theme)
                   It was about two in the morning and I was waiting for
                   the lights when a foreign looking woman jumped into the
                   car. "Drive", she said. My foot hit the floor. Five
                   seconds later all hell let loose, soldiers were
                   everywhere, tracker dogs, helicopters and armoured cars.
                   I saw a rifle pointed at the windscreen. She grabbed me
                   and literally threw me under the dashboard. There was a
                   sharp crack and the windscreen gave in. Moments later, I
                   was alone in the darkness – she'd gone, so had everyone
                   else. On the seat I saw a card. It read simply, 'silver
                   shield', they were with me in minutes.
                   See – I said you wouldn't believe me.
VOICE 2:           For the silver shield 24 hour windscreen service just
                   dial one hundred and ask for Freephone Silver Shield –
                   because you never know when you might need us.
                   (Music: up to finish)
[Courtesy County Sound Radio]                              (50 secs)
```

In a few seconds of airtime the script must gain our interest, make the
key point about the product (in the above case, immediacy) and say
clearly what action the listener must take to obtain it. This is especially
important for station-produced marketing promotions aimed at potential
buyers of its own advertising spots. Here the voices wittily imitate two
well-known cricket commentators:

```
FX:          Cricket atmosphere (held under throughout)

VOICE 1:     Yes, hello everyone and welcome to Southern Sounds small-ads
             county classic. It's a marvellous day here and interestingly
             enough, it's 9-99 for 5 transmissions. Quite an incredible
             offer – how did we arrive at that, John?

VOICE 2:     Well, it's 9-99 for 5 transmissions, once a night for a week.
             And that's the best small advertising offer in county radio in
             England since 1893.

VOICE 1:     Quite amazing, and all we have to do is call Alison small-ads,
             during playing hours on Brighton 4 triple 2 double 8. (Fx:
             light applause) And here comes an advertiser now – running in
             with his cheque for 9-99 (Fx: bat on ball, applause) – and it's
             superbly fielded by Alison small-ads – and it's on the air in a
             flash – very good effort that, I thought. And so with the offer
             still at 9-99 for 5 it's back to the studio.

                                                          (55 secs)
[Courtesy Southern Sound Radio – station marketing promo]
```

Advertising based on familiar radio programmes obviously strikes a
chord with the listener, given the right placing. The next example also
imitates BBC commentators, the writing parodying the sports style. This
is one of a series of ads making use of this device:

MAN 1:   So you want to be a football commentator, eh?

MAN 2:   Over the moon, Brian.

MAN 1:   Right, well you've gotta have all your football cliches -

MAN 2:   ah ha.

MAN 1:   plenty of drama -

MAN 2:   I'm way behind you there.

MAN 1:   and put your emphasis on - all the wrong words.

MAN 2:   I'm sure I <u>can</u>, manage that.

MAN 1:   Right - I'll give you a bit of the old crowd.

FX: <u>Football crowd noise (held under)</u>

MAN 2:   (in football commentary style) And we go into the second half
           with the score standing at 1-nil. So the game really is
           perfectly balanced - and I'm not going to sit on the fence but
           this game could go either way. You could take the atmosphere
           and cut it up in a thousand pieces, dip it in custard and give
           it to the crowd. And there's Gray, - Aston Villa's vibrant
           virtuoso who's decided to take his nerve by the horns and stamp
           his authority on this game in those Nike boots of his . . .

MAN 1:   (interrupting) Er - Hang on, hang on. <u>(Fx cut)</u> What did you say?

MAN 2:   Er Gray - Andy Gray the footballer.

MAN 1:   No, no - you said Nike, I'm sure I heard you say Nike.

MAN 2:   Quite categorically, yes.

MAN 1:   Yes but you can't mention brand names - OK? I mean what do you
           think this is - a commercial or something?

                                                           (60 secs)

<u>[Courtesy Grierson Cockman Craig & Druiff Ltd]</u>

This approach, once the format, style, and characters are established,
is very effective in promoting the brand name. The time may come when
for a national product the name need not be mentioned at all. The advantage
is that the listener joins in the game and is almost certain to say the
product name to himself – as in the famous 'Schhhh . . . you know who'
campaign. Here, Duracell batteries use only one voice – the same somewhat
tired, 'ordinary', older man's voice which was used for a time in all their
radio ads, plus their 'sound logo', used on TV as well:

VOICE: As an ordinary HP-8 grade radio battery I have one great
           ambition - I want to be forgotten. If I'm remembered it means
           I'm dead. Admit it - that's the only time you ever remember your
           batteries. You never ask us how we are or take us out for a
           nice walk - only the walk to the bin. Well you can forget me for
           145 hours of continuous radio noise. But there's a radio
           battery that can be forgotten for over 500 hours. Now that's
           what I call forgettable.

           You know the one - erm - oh well, wassisname.

```
ANNCR:      Wassisname (logo Fx: big door slam)
            No ordinary battery looks like it or lasts like it.
```

(50 secs)

[Courtesy Dorland Advertising Ltd]

Voicing and treatment

Casting a commercial is a make or break business. Doing it well stems
from having a clear idea of the overall impression required. Professional
actors may be expensive but are much more likely than the station's
office staff or the client firm's MD to provide what is required. They have
greater flexibility, vocal range and, above all, are 'produceable'. Once
they know what you want, trained performers will produce consistent
results – and you will certainly need it done many times, if only to get the
inflections, timing – with effects and music – absolutely right.

   Here's a striking piece of no-nonsense writing without music or effects.
The one voice has to be quiet, strong, tough and compassionate. It mustn't
be associated with any single professional or social group. The speed of
delivery is important.

```
MAN'S VOICE:    Our organisation currently has vacancies for people in
                this area.

                Successful applicants will get no company car, no luncheon
                vouchers, no holidays, no bonuses, no expense account, no
                business lunches, no glamour, and - no salary.

                If you're interested, and can spare a few hours a week to
                be a Samaritan, give us a ring for more information. We're
                in the phone book.

                But please don't phone unless you're serious, - someone
                else may be trying to get through.
```

(30 secs)

[Courtesy Saatchi & Saatchi Compton Ltd]

   In production, vocal inflection, emphasis, pace and projection are
infinitely variable. Even though you start with a clear idea of how a piece
should sound, you may want to try it a number of ways. It's generally
worth asking your speaker what sounds right or comfortable to him or
her. Should it be confidential, and relaxed or more excited? What emotional
content is appropriate? Should he 'shout', literally, to call attention, for
example, to a new store opening.

   There are times, however, when the end result is not the product of
creative artifice and technique, but comes straight from reality. Here is
the very real voice of genuine anguish – not an actress with words written
for her – but the halting voice, careful and controlled, of a mother who
had been interviewed after her son was killed in a road accident:

WOMAN'S VOICE:	They were crossing the road again after getting off the bus - and this crazy car came from nowhere and just took Simon - and he was killed instantly apparently.
	And I just knew when she said - I'm sorry, will you come and sit down? (voice breaking) I remember it so vividly.
	When the police came and told us that he had been - erm - charged - they just told us what an awful state he was in, and - erm - couldn't sleep, and was having nightmares, and he had a little boy of his own, aged five.
	The best way I can describe it is that when I thought about what had happened, I was - I preferred to be me than to be him. Because I didn't think I could live myself with the idea that I'd killed a small child.
MAN'S VOICE:	Drinking and driving wrecks lives.

(60 secs)

[Courtesy: COI/Dept of Transport, DMB&B]

The serious public service commercial such as this in particular needs the right voice – one that registers immediately to have impact. The presentation must reinforce the content. Many countries run advertising about the problems of AIDS and sexually transmitted disease – sensitive to the cultural context. Here's an award-winning British example from the Health Education Authority.

Music	Jingle Bells - non-vocal (establish and hold under)
Woman 1 (friendly and brisk)	If you're wondering what to give your loved one this Christmas, the following gifts are always worth considering - Chlamydia, genital warts, or even gonorrhoea. At this time of year these little surprises are as common as ever, and although you may know of the need to protect against HIV, these other infections can be very serious despite sometimes showing no obvious signs. Using a condom of course can help prevent their spread. So whatever you do give your partner for Christmas, make sure it's properly wrapped.
Music	(fade out)
Woman 2 (formal and authoritative)	For more information about HIV or other sexually transmitted infections, call the National AIDS helpline on 0800 567 123

(40 secs)

[Courtesy: BMP DDB Ltd]

The producer has a range of techniques to alter a voice; filters, digital effects, graphic equaliser and 'presence' control to change the tonal quality or to give a more incisive cutting edge, compression to restrict the dynamic range and keep the levels up, multi-tracking and variable-speed recording to increase the apparent number of voices, echo for 'space', phasing effects for mystery, and so on. Beware of gimmicks, however. If one is tempted to use technical tricks to 'make it more interesting', check the writing again – are the words really doing their job? One of the most effective ads I ever made was also the simplest. For a station in the tropics, its purpose was to sell a new iced lolly on a stick:

```
VOICE 1 (calling off):       Hey, where are you going?

VOICE 2 (calling, closer):   To get a Blums ice block.

ANNCR:                       Available in the best stores.
                                               (5 secs)
```

Run fairly frequently, because of its low cost, the phrases – or versions of them – could soon be heard all over town. An ad has to be right for its own culture. No matter how clever or complicated an advertisement is, it is never good on its own – only in relation to what happens after it is broadcast.

Music and effects

The main role of music is to assist in establishing mood. The biggest trap is to use a track from the library simply because of its title. The label may *say* 'Sunrise Serenade' but does it *sound* like an early morning promise of a new day, or is it cold, menacing, or just nondescript? Music in the context of the radio commercial must do what you want – immediately. If in doubt, play your choice to a colleague and ask 'what does this remind you of?' On your own you can convince yourself of anything.

The right music will almost certainly not be the right length. If you want the music to finish at the end, rather than to be faded, some judicious editing will be needed under the speech.

This was certainly the case in a witty ad using Mozart to promote holidays in Jamaica – in a British context a sensible guess at the tastes of the socio-economic target group likely to be interested in this type of holiday:

```
FX:                 Orchestra tuning up. Concert hall atmosphere

ANNCR (quietly):    And conducting Mozart's symphony number 40 in G minor
                    – Arturo Barbizelli – looking tanned and fit from his
                    recent holiday in Jamaica.

FX:                 Applause, quietens

MUSIC:              Mozart symphony No. 40 (4 bars) then accompanied by
                    steel drums, calypso style (4 bars)
                    (Music under)
```

ANNCR:	He was only there for a fortnight!
	(Music up, alternating between classical and calypso style. Music under)
ANNCR:	Find out about Jamaica – the island that warms you through and through.
JAMAICAN VOICE:	Ring Jamaica Tourist Board on 01-493-9007.
	(Music up and faded out)

(50 secs)

[Courtesy Young and Rubicam Ltd]

If the budget will run to specially commissioned music, however simple, you can clearly make it do what you want. Even a small station ought to be able to offer clients the possibility of tailor-made music – perhaps using a local musician with a bank of synthesisers. It can make all the difference – as with this firm of solicitors singing their own song. A comic, and memorable, idea which certainly gives the impression of its being a lively and 'unstuffy' firm.

MUSIC:	Upbeat piano accompaniment
VOICE 1:	I'm Underhill
VOICE 2:	I'm Wilcock
VOICE 3:	and I'm Taylor
ALL:	pleased to meet you
VOICE 1:	we're solicitors
ALL:	and jolly proud of it
VOICE 2:	in Wolverhampton
ALL:	Waterloo Road –
VOICE 3:	If you've a problem
ALL:	then we can help you
	as only a solicitor can
VOICE 1:	so that's Underhill
VOICE 2:	and Wilcock
VOICE 3:	and Taylor
ALL:	solicitors in Wolverhampton's Waterloo Road

(20 secs)

[Courtesy Beacon Radio – station produced commercial]

A word of warning about adding new words to a well-known song. Make sure that the music is out of copyright, or at least obtain written clearance from the publisher before proceeding. Publishers can be very sensitive about parodies of the work they own. All published music used in advertisements should be cleared in the same way that the station deals with its other music. For ads and promos Performing Rights and Copyright

Societies will normally make special arrangements rather than insisting on details of individual use.

On a technical point, make sure that the music/speech mix is checked for audibility in both the mono and stereo versions. The point made earlier is crucial – it may not be satisfactory to broadcast the same mix, in terms of relative level, on stereo FM as on mono medium wave.

Sound effects, like music, have to make their point immediately and unambiguously. They are best used sparingly, unless the impression required is one of chaos or 'busyness'. The right atmosphere effect to set the scene, manipulated and added to at appropriate points in the script – as in the 'cricket' example earlier – works well. Producers should not be misled by the title of an effects track or what it *actually* is. It is only what it *sounds* like that matters. And again, there is an armoury of techniques, from filters to vari-speed reproduction, for altering the sound.

Stereo

Commercials that deliberately exploit the stereo effect are relatively rare. There is the British Airways example which intersperses phrases from their music logo – the Flower Duet from 'Lakmé' by Delibes – using a sitar on the left and a piano on the right, eventually bringing them together under the speech line –

```
'Listen – the world is closer than you think.
British Airways – the world's favourite airline'
```

```
                                                              (60 secs)
```

[Courtesy: M & C Saatchi Ltd]

Even more pointed is this advertisement for the SEAT Arosa car. It doesn't make much sense on medium wave mono because of the two voices apparently speaking together, but it's clearly written and nicely produced for FM stereo.

```
Music      (slow alternating string chords. Hold under throughout)

Man's voice

       (L + R)  Your brain works in a mysterious way.
       (R)                                      Creative and emotive impulses,
                                                dreams and desires are
                                                controlled by the right side
                                                of the brain.
       (L)       Practical thoughts and
                 decisions are made by the
                 left side of the brain.
       (L + R)  So which will help you decide on a new SEAT?
                 Switch your stereo to left or right speaker for the
                 practical or emotive argument now.
```

Women's voices

(L + R) The SEAT Arosa. Neat, classy, with a big car feel and finish. The SEAT Ibitha Fresca, eager and sporty, puts the emphasis on control, handling, and good old fashioned entertainment. And the Cooper sports series, any more aggressive and it would need a muzzle. Twice formula 2 world rally champions, low cost, and seriously hairy. You're going to enjoy this. SEAT

The SEAT Arosa. A sub seven thousand pound price, and an equipment list to shame some mid-range five door hatchbacks. The Arosa is remarkably good value. The SEAT Ibitha Fresca, economical motoring, plenty of room, a spacious boot, central locking, and stereo radio cassette. The SEAT Cooper sports series brings race proven capabilities to everyday driving. Practical, safe, economical. SEAT

Man (L + R)   SEAT

Man/Woman

(L) (R) Enjoy yourself   Enjoy yourself

Man (L + R) Call the SEAT hotline on 0500 22 22 22 for details of your nearest dealer.

(60 secs)

[Courtesy: Chrysalis Creative/Galaxy 101]

Humour in advertising

We all like to laugh, and there is a perfectly logical connection between our liking an advertisement because it makes us laugh and liking the product which it promotes. The brand name endears itself to us by being associated with something that is witty and amusing. But the danger is twofold – that if the joke is too good it may obscure or send-up the product, and that if it is not good enough it will not stand up to one hearing, let alone the repetition that radio gives. The answer lies in genuinely comic writing that does not rely on a single punch line, and in characterisation that may be overplayed but which is nevertheless credible. The good commercial has much in common with the successful cartoon drawing. Even so, exposure of such wit should be carefully regulated. It is probably best to create a series of vignettes in a given style, and intermix them across the spot times to give maximum variety. The listener will enjoy the new jokes as well as welcoming old favourites – further, he will recall the product long before the ad gets to it.

One fairly sure-fire situation is to parody a popular TV programme, such as one of the soap operas, choosing voices and accents which represent already very familiar characters, or at least stereotypes:

```
MUSIC:              Big orchestral introduction (faded under)

ANNCR:              Nancy Loofah has something on her mind in today's
                    episode of 'Washingtown' – the long-running Fairy toilet
                    soap-opera.
                    (Music faded out. Fx slow clock tick)

MAN:                So, Nancy, why did you want to see me?

WOMAN:              Well, Doctor Flannel, you know that I'm head of the
                    Soft-Soap Corporation.

MAN:                Of course.

WOMAN:              Well, everywhere I turn I seem to see my biggest competitor.

MAN (reassuring):   OK, well let's try a little test. I'll say a word – you
                    say the first thing that comes into your head.

WOMAN:              OK.

MAN:                Let's start with 'pure'.

WOMAN:              Fairy.

MAN:                Mild.

WOMAN:              Fairy.

MAN:                Gentle.

WOMAN:              Fairy.

MAN:                Longer-lasting.

WOMAN:              Fairy.

MAN (smiling):      Right, well I'd say you had a mild –

WOMAN:              Fairy.

MAN:                – a mild fairy fixation. It should pass.

WOMAN:              Oh, how long will it last?

MAN:                Well, if it were an ordinary soap, not long. But as it's
                    Fairy . . .

WOMAN (gently):     It's all right doctor, you can tell me.

MAN:                As it's Fairy soap, Mrs Loofah, I'm afraid it could go on
                    for ages. (Music faded up)

WOMAN (horrified):  Oh, no!

ANNCR:              Has Nancy really been brainwashed? Listen soon for
                    'Washingtown' – the long-running Fairy toilet
                    soap-opera. (Music ends)
                                                          (60 secs)
```

[Courtesy Saatchi & Saatchi Compton Ltd]

   Some of these 'one-minuters' are full-scale dramas in their own right.
As has been said in another chapter, if the station facilities and expertise
cannot properly undertake this kind of work, it is better to succeed with
something simpler than embark on the complexity of a major production
such as this:

(American voices)	Fx hiss of space circuit, bleep
MAN 1 (on filter):	OK base, I'm on the ladder now. (Music: majestic, low level) This is one small leap for man . . .
MAN 2:	Hank, er, just hold on there will you. (Music cuts) That's 'one small step for man' there.
MAN 1 (filter):	Er, yes sir. (To self) One small step – OK base. (Music begins) This is one tiny (music cuts) – oh rats.
MAN 2 (laconic):	Relax, Hank, just take your time now. (Music begins)
MAN 1 (filter):	This is one big, small, step here, for one man to take off a ladder . . . (music cuts)
MAN 2:	OK, Hank, you're getting warm. 'One giant leap for mankind'.
MAN 3 (filter):	Hank, I'm getting cramp on the ladder up here, will you hurry it up please. (Music begins)
MAN 1 (filter):	This is one small leap for a giant – (music cuts)
MAN 2 (interrupting):	One giant leap for mankind. (Music begins)
MAN 1 (filter):	This kind man is a small giant. (Music cuts)
MAN 2 (testily):	Giant leap. (Music begins)
MAN 1 (filter):	This leap year is gonna be the best (music cuts) – aah
MAN 2:	Chuck, will you unload the Heineken bay and refresh that man's speech please?
FX (close):	Pouring liquid into a glass
MAN 1 (filter):	OK, I'm ready – run the music. (Music begins again) This is one small step for man – one giant leap for mankind. Fx applause up Music: men's voices sing slowly in the style of 'Space Odyssey' 'Heineken refreshes the parts other beers cannot reach'.
MAN 1 (filter; triumphant):	The Blue Tit has landed.
MAN 2:	The Eagle, Hank, the Eagle.

(85 secs)

[Courtesy Lowe Howard-Spink Marschalk]

A radio commercial is trying to sell a real product or service: the advertisement itself must therefore have reality, however outrageous its style. In the end, people must believe it.

10

The discussion programme

The topic for a broadcast debate should be a matter in which there is genuine public interest or concern. The aim is for the listener to hear argument and counter-argument expressed in conversational form by people

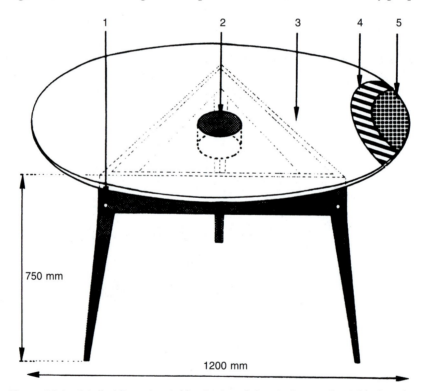

Figure 10.1 A talks/discussion table. Designed for studio use the table has three legs to reduce obstruction and to prevent it wobbling on an uneven floor. 1. Headphone jack carrying a programme feed with or without additional talkback. 2. Centre hole to take the microphone placed either on a special carrier fitting or on a floor stand. 3. Acoustically transparent table top consisting of a loose-weave cloth surface, a layer of padding (4) and a steel mesh base (5). (Courtesy of BBC Engineering Information Department)

actually holding those views with conviction. The broadcaster can then remain independent.

Format

In its simplest form there will be two speakers representing opposing views together with an impartial chairman. The producer may of course decide that such an arrangement would not do justice to the subject, that it is not as clear cut as the bidirectional discussion will allow and it might therefore be better to include a range of views – the 'multi-facet' discussion.

In this respect the 'blindness' of radio imposes its own limitations and four or five speakers should be regarded as the maximum. Even then it is preferable that there is a mix of male and female voices.

Under the heading of the discussion programme should also come what is often referred to as 'the chat show'. Here, a well-known radio personality introduces a guest and talks with him. It may incorrectly be described as an interview but since 'personalities' have views of their own which they are generally only too ready to express, the result is likely to be a discussion. The 'chairman plus one' formula can be a satisfactory approach to a discussion, particularly with the more lightweight entertainment, the non-current-affairs type of subjects. It works less well in the controversial, political, current-affairs field since it is more difficult

Figure 10.2 A discussion can become confusing if it contains too many different points of view

for the chairman to remain neutral if he is part of the discussion. In any case the danger for the broadcaster is that in order to draw out his guest contributor, he is always acting as 'the opposition' and becomes identified as 'anti-everything'. In such cases the more acceptable format is that of the interview. It is important for listeners and broadcasters to draw a clear distinction between an interview and a discussion.

A very acceptable format is where questions are put – possibly from an audience – to a panel of speakers representing different political parties or specialist interests. Such 'Any Questions?' programmes have a sharper edge by being broadcast live.

Selection of participants

Do all participants start equal? A discussion tends to favour the articulate and well organised. The chairman may have to create opportunities for others to make their case. It is possible to 'load' a discussion so that it is favourable to a particular point of view, but since the listener must be able to make up his mind by hearing different views adequately expressed, the producer should look for balance – of ability as well as opinion. Often there are, on the one hand, the 'official spokesmen', and, on the other, the good broadcasters! Sometimes they combine in the same person but not everyone, in the circumstances of the broadcast debate, is quick thinking, articulate and convincing – however worthy they may be in other respects. In selecting the spokesman for one political party it is virtually obligatory also to include his 'shadow' opposite number – whatever the quality of his radio performance.

There will obviously be times when it is necessary to choose the leader of the party, the council member, the chairman of the company or the official spokesperson; but there will also be occasions when the choice is more open to the broadcaster and in the multi-faceted discussion it is important to include as diverse a range of interests as is appropriate.

In general terms these can be summarised as: power holders and decision makers, legal representation and 'watch-dog' organisations, producers of goods and services, and consumers of goods and services. These categories will apply whether the issue is the flooding of a valley for a proposed hydro-electric scheme, a change in the abortion law, or an increase in the price of food.

The listener should also be regarded as a participant and the topic should at least be one which involves him. If the listener is directly affected he can be invited to take part in a follow-up programme either by letter, phone, fax or e-mail. In the event of a public meeting on the subject it may be that the broadcaster can arrange to cover it with an outside broadcast.

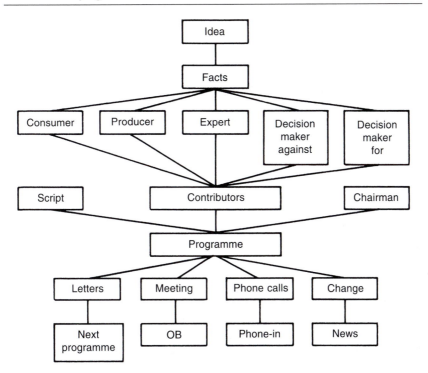

Figure 10.3 Stages in producing a discussion programme. Select the topic – research the information – choose the participants – coordinate the contributors – broadcast the programme – deal with the response – evaluate the possibility of follow-up

The chairperson

Having selected the topic and the team, the programme will need someone to chair the discussion. The ideal is knowledgeable, firm, sensitive, quick thinking, impartial, and courteous. He or she will be interested in almost everything and will need a sense of humour – no mean task!

Having obtained this paragon of human virtue who also possesses a good radio voice and an acute sense of timing, there are several points which need attention before the broadcast. One of these is to decide what to call the chairperson. Here the term chairman is used to include both men and women.

Preparation

The subject must be researched and the essential background information gathered and checked. Appropriate reference material may be found in libraries, files of newspaper cuttings, on the Internet, and in the radio station's own newsroom. The chairman must have the facts at his fingertips and have a note of the views already expressed so that he understands the points of controversy. He can then prepare a basic 'plot' of the discussion outlining the main areas to be covered. This is in no sense a script, it is a reminder of the essential issues in case they should get sidetracked in the debate.

It is important that the speakers are properly briefed beforehand making sure that they understand the purpose, range *and limitations* of the discussion. They should each know who is to take part and the duration of the programme. It is not necessary for them to meet before the broadcast but they should be given the opportunity to do their own preparation.

Starting the programme

At the start of the broadcast the chairman introduces the subject making it interesting and relevant to the listener. This is often done by putting a series of questions on the central issues, or by quoting remarks already made publicly.

The chairman should have everyone's name, and his or her designation, written down in front of him. He then introduces them, making sure that all their voices are heard as early as possible in the programme. During the discussion he should continue to address them by name, at least for the first two 'rounds' of conversation. Their names should be used again at intervals throughout. It is essential that the start of the programme is factual in content and positive in presentation. Such an approach will be helpful to the less confident members of the team and will reassure the listener that the subject is in good hands. It also enables the participants to have something 'to bite on' immediately so that the discussion can begin without a lengthy warming-up period.

Speaker control

In the rather special conditions of a studio discussion, some people become highly talkative believing that they have failed unless they have put their whole case in the first five minutes. On the other hand, there are the nervously diffident. It is not possible to make a poor speaker appear brilliant, but there is an important difference between someone with poor delivery and someone with little to say. The chairman must draw out the former and curb the latter. Even the most voluble have to breathe – a factor which repays close observation! The chairperson's main task is to provide equal opportunity of expression for all participants. To do this may require suppression as well as encouragement, and such disciplines as are required should be communicated – probably non-verbally.

After having an opinion strongly expressed, that speaker should not be allowed to continue for too long before another view of the matter is introduced. The chairman can interrupt, and it's best done constructively . . .

'That's an important point, before we go on, how do others react to that? – Mrs Jones?' The chair must in these cases give a positive indication by voice, facial expressions and possibly hand signal as well of who is to speak. It is also necessary to prevent two voices from speaking at once, other than for a brief interjection, by a decisive and clear indication of 'who holds the floor'. It is not a disaster when there are two or more voices, indeed it may be a useful indicator of the strength of feeling. It has to be remembered, however, particularly when broadcasting monophonically, that when voices overlay each other, the listener is unlikely to make much sense of the actual content.

Subject control

The chairman has to obtain clarification of any technical jargon or specialist language which a contributor may use. Abbreviations, particularly of organisations, are generally far less well understood by the listener than people sitting round a studio table would like to think.

With one eye on his prepared 'plot' and the other on the clock, the chairman steers the subject through its essential areas. However, it's important to remain reasonably flexible and if one particular aspect is proving especially interesting, the chair may decide to depart from the original outline. Questions in the chairman's mind should be:

- Time gone – time to go.
- How long has he had?
- Is it irrelevant?
- Is it boring?
- Is it incomprehensible?
- Next question.
- Who next?

Above all, he must be able to spot and deal with red herrings and digressions. To do this he must know where the discussion should be going and have the appropriate question phrased so that he can interrupt positively, constructively and courteously.

In a lengthy programme it may be useful to introduce a device which creates variety and helps the discussion to change direction. Examples are a letter from a listener, or quote from an article read by the chairman, a pre-recorded interview, a piece of actuality, or a phone call. If the chairman is to retain his impartiality such an insert should not be used to make a specific point but simply to raise questions on which the participants may then comment.

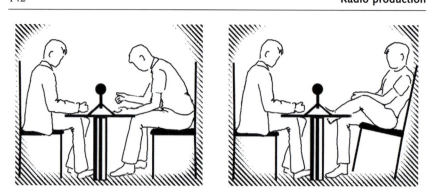

Figure 10.4 Voice levels must be watched throughout. A person with a quiet voice will have to sit close in to the table and the discussion chairman must prevent too much movement

Technical control

The chairman has to watch for, and correct, alterations in the balance of voices which was obtained before the programme began. This may be due to a speaker moving back 'off-mic', turning directly to address his neighbour or leaning in too close. There may be wide variations in individual voice levels as the participants get annoyed, excited, discomfited or subdued. And he has to be aware of any extraneous noise such as paper rustle, matches being struck or fingers tapping the table. Non-verbal signals should suffice to prevent them becoming too intrusive.

As an aid in judging the effect of any movement, changes in voice level or unwanted sounds, the chairman will often wear headphones. These should be on one ear only to avoid his being isolated acoustically from the actual discussion. This will also enable him to hear talkback from the producer so that he can be fed with additional ideas – for example, on a point of the discussion which might otherwise be overlooked. On occasions, everyone in the studio will require headphones. This is likely if the programme is to include phone calls from listeners, or when members of the discussion group are not physically present in the same studio but are talking together over links between separate studios. In these circumstances, the talkback arrangements have to be such that the producer's editorial comments are confined only to the headphones worn by the chairman. To avoid embarrassment and confusion such a system must be checked before the programme begins.

An important part of the technical control of the programme is its overall timing. The chairman must never forget the clock.

Ending the programme

It is rarely desirable for the chairman to attempt a summing-up. If the

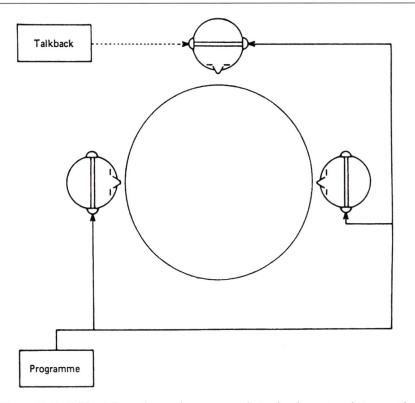

Figure 10.5 Talkback from the producer goes only to the discussion chairman. The other participants may also have to wear headphones, carrying the programme feed only, in order to hear a remote contributor or phone call

discussion has gone well, the listener will already have recognised the main points being made and the arguments which support them. If a summary *is* required, it is often better to invite each speaker to have a 'last word'. Alternatively, the chairman may put a key question to the group which points the subject forward to the next step – 'Finally, what do you think should happen now?' This should be timed to allow for sufficient answering discussion.

Many a good programme is spoiled by an untidy ending. The chairman should avoid giving the impression that the programme simply ran out of time:

'Well, I'm afraid we'll have to stop there . . .'
'Once again the old clock has beaten us . . .'
'What a pity there's no time to explore that last point . . .'

The programme should cover the material which it intended, in the time it was allowed. With a minute or less to go, the chairman should thank the contributors by name, giving any other credits due and referring to further programmes or public events related to the subject.

After the broadcast comes the time when the participants think of the remarks they should have made. An opportunity for them to relax and 'unwind' is important, and this is preferably done as a group, assuming they are still speaking to each other. They are at this stage probably feeling vulnerable and exposed, wondering if they have done justice to the arguments they represent. They should be warmly thanked and allowed to talk informally if they wish. The provision of some refreshment or hospitality is often appropriate.

It is not the broadcaster's job to create confrontation and dissent where none exists. But genuine differences of opinion on matters of public interest offer absorbing broadcasting since the listener may feel a personal involvement in the arguments expressed and in their outcome. The discussion programme is a contribution to the wider area of public debate and may be regarded as part of the broadcaster's positive role in a democratic society.

11

The phone-in programme

Critics of the phone-in describe it as no more than a cheap way of filling air-time and undoubtedly it is sometimes used as such. But like anything else, the priority it is accorded and the production methods applied to it will decide whether it is simply transmitter fodder or whether it can be useful and interesting to the listener. 'It's Your World' on the BBC's World Service has at least the potential for putting anyone anywhere in touch with a major international figure to question policy and discuss issues of the day.

Through public participation the aim of a phone-in is to allow a democratic expression of view and to create the possibility of community action. An important question therefore is to what extent such a programme excludes those listeners who are without a telephone. Telephone ownership can vary widely between regions of the same country, and the cities are generally far better served than the rural areas. Per head of population North America has more telephones than anywhere else in the world. It follows that to base programmes simply on the American or Canadian practice may be misleading. It is a salutary reminder that 50 per cent of the world's population has never used a phone.

It is possible to be over-glib with the invitation – '. . . if you want to take part in the programme just give us a ring on . . .' Cannot someone take part simply by listening? Or to go further, if the aim is public participation, will the programme also accept letters, or people who actually arrive on the station's doorstep? The little group of people gathered round the door of an up-country studio in Haiti, while their messages and points of view were relayed inside, remains an abiding memory of a station doing its job.

It is especially gratifying to have someone, without a phone at home, go to the trouble of phoning from a public call box. To avoid losing the call when the money runs out, the station should always take the number and initiate such calls as are broadcast on a phone-back basis.

Technical facilities

When inviting listeners to phone the programme it is best to have a special number rather than take the calls through the normal station telephone number. Otherwise the programme can bring the general telephone traffic to a halt. The technical means of taking calls have almost infinite variation but the facilities should include:

1   Off-air answering of calls.
2   Acceptance of several calls – say four or five simultaneously.
3   Holding a call until required, sending it a feed of cue programme.
4   The ability to take two calls simultaneously on the air.
5   Origination of calls from the studio.
6   Picking up a call by the answering position after its on-air use.

Programme classification

The producer of a phone-in must decide the aim of the programme and design it so that it achieves a particular objective. If he simply throws the phone lines open to listeners, the result can be a hopeless muddle. There are always cranks and exhibitionists ready to talk without saying anything, and there are the lonely with a real need to talk. Inexpert advice given in the studio will annoy those listeners who know more about the subject than the presenter, and may actually be harmful to the person putting the question. It is essential that the producer knows what he is trying to do and by an adequate screening of the incoming calls, limits the public participation to the central purpose of the programme.

Types of phone-in include:

1   The open line – conversation with the studio presenter.
2   The specific subject – expert advice on a chosen topic.
3   Consumer affairs – a series providing 'action' advice on detailed cases.
4   Personal counselling – problems discussed for the individual rather than the audience.

The open line

A general programme where topics of a non-specific nature are discussed with the host in the studio. There need be no theme or continuity between the calls but often a discussion will develop on a matter of topical interest. The 1-minute phone-in, or 'soap-box', works well when callers are allowed to talk on their own subject for one minute without interruption, providing they stay within the law.

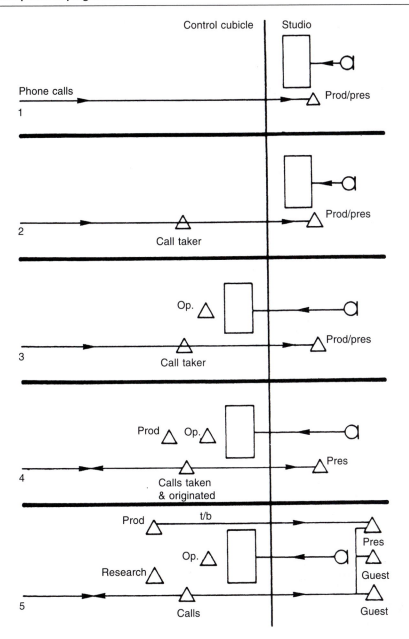

Figure 11.1 Staffing a phone-in. 1. The total self-op. The presenter takes his own calls. 2. A 'call taker' screens the calls and provides information to the presenter in advance of each call. 3. Operator to control all technical operation, e.g. discs, levels, etc., while producer/presenter concentrates on programme content. 4. Separate producer in control area to make programme decisions, e.g. to initiate 'phone-out' calls. 5. Guest 'experts' in the studio with research support available

Support staff

There are several variations on the basic format in which the presenter himself simply takes the calls as they come in. The first of these is that the lines are answered by a programme assistant or secretary who ensures that the caller is sensible and has something interesting to say. The assistant outlines the procedure – 'please make sure your radio set is not turned on in the background' (to avoid acoustic 'howl-round') – 'you'll hear the programme on this phone line and in a moment X (the presenter) will be talking to you'. The call is held until the presenter wants to take it. Meanwhile the assistant has written down the details of the caller's name and the point he wishes to make and this is passed to the presenter. Since they are in separate rooms studio talkback will be used or, better, a third person involved to pass this information. This person is often the producer who will decide whether or not to reject the call on editorial or other grounds, to take calls in a particular order, or to advise the presenter on how individual calls should be handled. If the staffing of the programme is limited to two, the producer should take the calls since this is where the first editorial judgement is made. A computer is useful for providing visual talkback between call-taker and presenter, so that the presenter can be given a considerable amount of advance information about each call.

Choosing the calls

The person vetting the calls quickly develops an ear for the genuine problem, the interesting point of view, the practical, or the humorous. Such people *converse*. They have something to say but can listen as well as speak, they tend to talk in short sentences and respond quickly to questions put to them. These things are soon discovered in the initial off-air conversation. Similarly there are people who one might prefer not to have on the programme:

- the 'regulars' who are always phoning in
- the abusive, perverted, offensive, or threatening caller
- the overtalkative and uninterruptible
- the boring, dull or slow
- those with a speech defect, unless the programme is specifically designed for this topic
- those with speech patterns or accents which make intelligibility particularly difficult
- the sycophantic, who only want to hear you say their name on the air.

Of course the programme must have the idiosyncratic as well as the 'normal', and perhaps anger as well as conciliation. But without being too rude there are many ways of ending a conversation, either on or off the air:

'I'm sorry we had someone making that point yesterday . . .'
'I can't promise to use your call, it depends how the programme goes . . .'
'There's no one here who can help you with that one . . .'
'I'm afraid we are not dealing with that today . . .'
'This is a very bad line, I can't hear you . . .'
'You are getting off the subject, we shall have to move on . . .'
'We have a lot of calls waiting . . .'

or simply

'Right, thank you for that – Goodbye . . .'

The good presenter will have the skill to turn someone away without turning them off, but in moments of desperation it is worth remembering that even the most loquacious person has to breathe.

The role of the presenter

The primary purpose of the programme is democratic – to let people have their say and express their views on matters which concern them. It is equivalent to the 'letter to the editor' column of a newspaper or the soapbox orator stand in the city square. The role of the presenter or host is not to take sides – although some radio stations may adopt a positive editorial policy – it is to stimulate conversation so that the matter is made interesting for the listener. He must be well versed in the law of libel and defamation and be ready to terminate a caller who becomes obscene, overtly political, commercial or illegal in accordance with the programme policy.

Very often such a programme succeeds or fails by the personality of the host presenter – quick thinking with a broad general knowledge, interested in people, well versed in current affairs, wise, witty, and by turn as the occasion demands genial, sharp, gentle, possibly even rude. All this combined with a good characteristic radio voice, the presenter is a paragon of broadcasting.

Reference material

The presenter may be faced with a caller actually seeking practical advice and it is important for the producer to know in advance how far the programme should go in this direction, otherwise it may assume expectations for the listener which cannot be fulfilled. Broadcasters are seldom recruited for their practical expertise outside the medium and there is no reason why they should be expected spontaneously to answer specialist questions. However, the availability in the studio of suitable reference material will enable the presenter to direct the caller to the appropriate source of advice or information. Reference sources may include telephone directories, names and addresses of councillors, members of parliament or other elected representatives, government offices, public utilities, social services, welfare organisations and commercial PR departments. This information is usually given on the air; but it is a matter of discretion. In certain cases it may be preferable for the presenter to hand the caller back to the secretary who will provide the appropriate information individually. If there is a great deal of factual material needed then a fourth person will be required to do the immediate research, especially if searching the Internet.

Studio operation

At the basic level it is possible for the presenter himself to undertake the operation of the studio control desk. But as facilities are added, it becomes necessary to have a specialist panel operator, particularly where there is no automatic equipment to control the sound levels of the different sources. In this respect an automatic 'voice-over' unit for the presenter is particularly useful, so that when he speaks the level of the incoming call is decreased. It must, however, be used with care if he is to avoid sounding too dominating.

Additional telephone facilities

If the equipment allows, the presenter may be able to take two calls simultaneously so setting up a discussion between callers as well as with himself. The advice and cooperation of the telephone company is required prior to the initiation of any phone programme. This is because there may well have to be safeguards taken to prevent the broadcasting function from interfering with the smooth running of the telephone service. These may take the form of limitations imposed on the broadcaster in how he may use the public telephone in programmes, or possibly the installation of special equipment either at the telephone exchange or at the radio station.

Use of 'delay'

The listening interest of this type of programme depends to an extent on the random nature of the topics discussed and the consequent possibility of the unexpected or outrageous. There is a vicarious pleasure to be obtained from a programme not wholly designed in advance. But it is up to the presenter to ensure that there is reasonable control. However, as an additional safeguard, it is possible to introduce a delay time between the

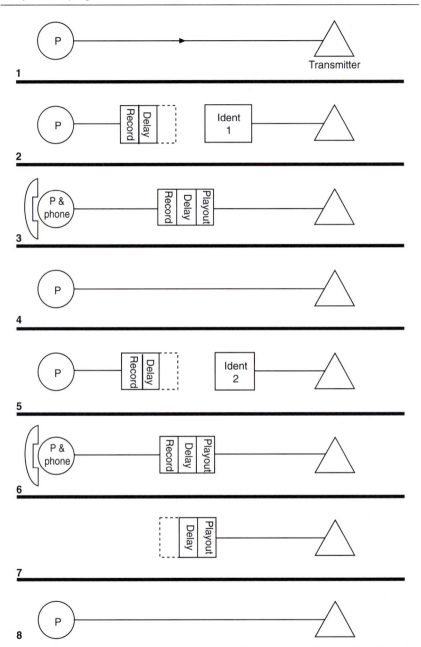

Figure 11.2 The use of delay in a phone-in. **1** The presenter P is fed directly to the transmitter. **2** To introduce a delay the presenter is recorded and held on a short-term – 10 seconds – storage. The programme output is maintained for this duration by Ident 1. **3** As Ident 1 ends, the programme continues using the recorded playout. The transmission is now 10 seconds behind 'reality'. **4** If a caller says something that has to be cut, the delayed output is replaced by the presenter 'live'. **5** To reinstate the delay the procedure is as 2, but using a different Ident. **6** The programme continues through the delay device. **7** The presenter brings the programme to an end allowing the recorded playout to finish the transmission. **8** Normal live presentation

programme and the transmission – indeed some radio stations and broadcasting authorities insist on it. Should any caller become libellous, abusive or obscene a delay device, of say ten seconds, enables that part of the programme to be deleted before it goes on the air. The programme, usually a short-term recording, is faded out and is replaced by the 'live' presenter's voice. With a good operator, this substitution can be made without it being apparent to the listener. Returning from the 'live' to the 'delayed' programme is more difficult and it is useful to have on hand news, music or other breaks to allow the presenter time to return to another call. If the caller is occasionally using words which the producer regards as offensive these can be 'bleeped out' by replacing them with a tone source as they reach the air. Again, good operation is essential. Overall, however, a radio station gets the calls it deserves and given an adequate but not oppressive screening process, the calls will in general reflect the level of responsibility at which the programme itself is conducted.

The specific subject

Here, the subject of the programme is selected in advance so that the appropriate guest expert, or panel of experts, can be invited to take part. It may be that the subject lends itself to the giving of factual advice to individual questions, for example child care, motoring, medical problems, gardening, pets and animals, antiques, holidays, cooking, citizens' rights. Or the programme may be used as an opportunity to develop a public discussion of a more philosophical nature, for instance the state of the economy, political attitudes, education or religious belief.

A word of warning, however, about giving specific advice.

Doctors should not attempt diagnosis over the phone, much less prescribe drugs in individual cases. What tends to be most useful is general information about illness, the side-effects of drugs, what to expect when you go into hospital, the stages of pregnancy, children's diseases, reclaiming medical costs on social welfare, and so on.

Of course there will be difficult calls and the producer should discuss his programme policy beforehand with his guests rather than having to decide it live on-air. For example, how will you deal with cancer, AIDS and the terminally ill? At what point might you rule out a call as being too embarrassing, disgusting, or voyeuristic? When is the caller just sending you up?

The legal phone-in too can provide a wide range of useful subjects rooted in actual practice; employment law, motor insurance claims, marriage and divorce, wills, the rights of victims, disputes between landlord and tenant, land claims, house purchase, the responsibilities of children, arguments between neighbours – noise, nuisance, fences, trees, etc. A lawyer, like the doctor, should be wary of providing specific solutions over the phone – the caller is unlikely to give all the facts, especially those not supporting his or her case. There is also a real danger of being used as a second opinion in a case already in front of the courts, or where the caller is dissatisfied with the advice of his own lawyer. The professional guest will know when to respond with: 'I can't comment on your actual case without the details, but in general . . .' The point of whether a radio service can be sued for giving wrong advice is of interest here. The answer will be 'no', so long as the programme does not try to imitate the private consultation. The producer serves the public best by providing advice of general applicability, illustrating how the law operates in specific areas.

'Early lines'

In order to obtain questions of the right type and quality, the phone lines to the programme may be opened some time before the start of the transmission – say half an hour. The calls are taken by a secretary or programme assistant who notes the necessary details and passes the information to the producer who can then select the calls he wants for the programme. For the broadcast these are originated by the studio on a phone-back basis.

The combination of 'early' lines and 'phone-back' gives the programme the following advantages:

1   The calls used are not random but are selected to develop the chosen theme at a level appropriate to the answering panel and the aim of the programme.
2   The order in which the calls are broadcast is under the control of the producer and so can represent a logical progression of the subject.
3   The studio expert, or panel, has advance warning of the questions and can prepare more substantial replies.
4   The phone-back principle helps to establish the credentials of the caller and serves as a deterrent to irresponsible calls. The programme itself may therefore be broadcast 'live' without the use of any delay device.
5   At the beginning of the programme there is no waiting for the first calls to come in, it can start with a call of strong general interest already established.
6   Poor or noisy lines can be redialled by the studio until a better quality line is obtained.

Consumer affairs

The consumer phone-in is related to the 'specific subject' but its range of content is so wide that any single panel or expert is unlikely to provide detailed advice in response to every enquiry. As the range of programme content increases, the type of advice given tends to become more general, dealing with matters of principle rather than the action to be taken in a specific case. For example, a caller complains that an electrical appliance bought recently has given persistent trouble – what should they do about it? An expert on consumer legislation in the studio will be able to help distinguish between the manufacturer's and the retailer's responsibility, or whether the matter should be taken up with a particular complaints council, electricity authority or local government department. To provide a detailed answer specific to this case requires more information. What were the exact conditions of sale? Is there a guarantee period? Is there a service contract? Was the appliance being used correctly?

The need to be fair

Consumer affairs programmes rightly tend to be on the side of the complainant, but it should never be forgotten that a large number of complaints disintegrate under scrutiny and it is possible that such fault as there is lies with the user. Championing 'the little man' is all very well, but radio stations have a responsibility to shopkeepers and manufacturers too. Once involved in a specific case the programme must be fair, and be seen to be fair. Two further variations on the phone-in help to provide this balance:

1   *The phone-out.* A useful facility while taking a call is to be able to originate a second call and have them both on the air simultaneously. In response to a particular enquiry, the studio rings the appropriate head of sales, PR department, or council/government official to obtain a detailed answer, or at least an undertaking that the matter will be looked into.
2   *The running story.* The responsibility to be fair often needs more information than the original caller can give or than is immediately available and an enquiry may need further investigation outside the programme. While the problem can be posed and discussed initially, it may be that the subject is one which has to be followed up later.

Linking programmes together

Unlike the 'specific subject' programme which is an individual 'one-off', the broad consumer affairs programme may run in series – weekly, daily or even morning and afternoon. A complex enquiry may run over several programmes and while it can be expensive of the station's resources, it can also be excellent for retaining and increasing the audience. For the long-term benefit of the community, and the radio station, the broadcaster must assure himself that such an investigation is performing a genuine public service.

As with all phone-in programmes consideration must be given to recording the broadcast output in its entirety as a check on what was said – indeed this may be a statutory requirement imposed on the broadcaster. People connected with a firm which was mentioned but who heard of the broadcast at second-hand may have been given an exaggerated account of what transpired. An ROT (recording off transmission) enables the station to provide a transcript of the programme and is a wise precaution against allegations of unfair treatment or threat of legal proceedings – always assuming of course that the station has been fair and responsible!

Personal counselling

With all phone-in programmes, the studio presenter is talking to the individual caller but has constantly to bear in mind the needs of the general listener. The material discussed has to be of interest to the very much wider audience who might never phone the station but who will identify with the points raised by those who do. This is the nature of broadcasting. However, once the broadcaster declares that he will tackle personal problems, sometimes at a deeply psychological and emotionally disturbing level, he cannot afford to be other than totally concerned for the welfare of the individual caller. For the duration, his responsibilities to the one exceed those to the many. Certainly, the presenter cannot terminate a conversation simply because it has ceased to be interesting or because it has become too difficult. Once a radio service says 'bring your problems here', it must be prepared to supply some answers.

This raises important questions for the broadcaster. Is he exploiting individual problems for public entertainment? Is the radio station simply providing the opportunities for the aural equivalent of the voyeur? Or is there sufficient justification in the assumption that without even considering the general audience, at least some people will identify with any given problem and so be helped by the discussion intended for the individual? It depends of course on how the programme is handled, the level of advice offered, and whether there is a genuine attempt at 'caring'.

To what extent will the programme provide help outside its own air time? The broadcaster cannot say 'only bring your depressive states to me between the transmission times of 9 to 11 at night'. Having offered help, what happens if the station gets a call of desperation during a music show? The radio station has become more than simply a means of putting out programmes, it has developed into a community focus to which people turn in times of personal trouble. The station must not undertake such a role lightly, and it must have sufficient contacts with community services that it can call on specialist help to take over a problem which it cannot deal with itself. But how can the Programme Director ensure that the advice given is responsible? Because of all the types of programme which a station puts out, this is the one where real damage may be done if the broadcaster gets it wrong. Discussing problems of loneliness, marriage, and sex, or the despair of a would-be suicide, has to be taken seriously. It is important to get the caller to talk and to enlist his support for the advice given, and for this purpose a station needs its trained counsellors.

The presenter as listener

As with all phone-in programmes the presenter in the studio cannot see the caller. He is denied all the usual non-verbal indicators of communication – facial expression, gesture, etc. This becomes particularly important in a counselling programme when the caller's reaction to the advice given is crucial. The person offering the advice must therefore be a perceptive listener – a pause, a slight hesitation in what the caller is saying may be enough to indicate whether he is describing a symptom or a cause, or whether he has yet got to the real problem at all. For this reason many such programmes will have two people in the studio – the presenter, who will take the call initially and discuss the nature of the problem – and the specialist counsellor who has been listening carefully and who takes over the discussion at whatever stage he feels necessary. Such a specialist may be an expert on marriage guidance, a psychiatrist, a minister of religion, or doctor.

Non-broadcasting effort

It is important that the programme also has off-air support – someone to talk further with the caller or to give names, addresses or phone numbers which are required to be kept confidential. The giving of a phone number over the air is always a signal for some people to call it, so blocking it as an effective source for the one person the programme is trying to help. Again, the broadcaster may need to be able to pass the problem to another agency for the appropriate follow-up.

The time of day for a broadcast of this type seems to be especially critical. It is particularly adult in its approach and is probably best at a time when it may be reasonably assumed that few children will be listening. This indicates a late evening slot – but not so late as to prevent the availability of unsuspected practical help arising from the audience.

Anonymity

Often a programme of this type specialising in personal problems allows the caller to remain unidentified. His name is not given over the air, the studio counsellor referring to him by his Christian name only or by an agreed pseudonym. This convention preserves what most callers need –

privacy. It is perhaps surprising that people will call a radio station for advice, rather than ask their family, friends or specialist, simply because they do not have to meet anyone. It can be done from a position of security, perhaps in familiar surroundings where they do not feel threatened.

People with a real problem seldom ring the station in order to parade it publicly – such exhibitionists should be weeded out in the off-air screening and helped in some other way. The genuine seeker of help calls the station because he already knows it as a friend and as a source of unbiased personal and private advice. He knows he need not act on that advice unless he agrees with it. This is a function unique to radio broadcasting. It is perhaps a sad comment when a caller says 'I've rung you because I can't talk to anyone about this', but it is in a sense a great compliment to the radio station to be regarded in this way. As such it must be accepted with responsibility and humility.

Phone-in checklist

The following list summarises what is needed for a phone-in programme:

1 Discuss the programme with the telephone service and resolve any problems caused by the additional traffic which the programme could generate. Do you want all the calls, even the unanswered ones, to be counted?
2 Decide the aim and type of the programme.
3 Decide the level of support staff required in the studio. This may involve a screening process, phone-back, immediate research, operational control and phone-out.
4 Engage guest speakers.
5 Assemble reference material.
6 Decide if 'delay' is to be used.
7 Arrange for 'Recording Off Transmission'.
8 Establish appropriate 'follow-up' links with other, outside, agencies.

12

The vox pop

'Vox populi' is the voice of the people, or 'man in the street' interview. The use of the opinions of 'ordinary' members of the public adds a useful dimension to the coverage of a topic which might otherwise be limited to a straight bulletin report or a studio discussion among officials or experts. The principle is for the broadcaster using a portable recorder to put one, possibly two, specific questions on a matter of public interest to people selected by chance, and to edit together their replies to form a distillation of the overall response. While the aim is to present a sample of public opinion, the broadcaster must never claim it to be statistically valid, or even properly representative. It can never be anything more than 'the opinions of some of the people we spoke to this afternoon'. This is because gathering material out on the streets for an afternoon magazine programme will almost certainly over-represent shoppers, tourists and the unemployed; and be low on businessmen, motorists, night shift workers and farmers! Since the interviewing is done at a specific time and generally at a single site, the sample is not really even random – it is merely unstructured and no one can tell what the views obtained actually represent. So no great claim should be made for the sample 'vox pop' on the basis of its being truly 'the voice of the people'.

It is easier to select a specific grouping appropriate to a particular topic – for example, early risers, commuters, children or lorry drivers. If the question is to do with an increase in petrol prices, one will find motorists, together with some fairly predictable comment, on any garage forecourt. Similarly, a question on medical care might be addressed to people coming out of a hospital. Incidentally, many apparently public places are in fact private property and the broadcaster must remember that he has no prescriptive right to work there without permission.

As the question to which reaction is required becomes more specific, the group among which the interviews are carried out may be said to be more representative. Views on a particular industrial dispute can be canvassed among the pickets at the factory gate, opinions on a new show sought among the first-night audience. Nevertheless it is important in the presentation of 'vox pop' material that the listener is told where and when it was gathered. There must be no weighting of the interview sample of which the listener is unaware. Thus the introductory sentence 'We asked the strikers themselves what they thought' may mislead by being more comprehensive than the actual truth. A more accurate statement would be: 'We asked some of the strikers assembled at the factory gate this morning what they thought'. It is longer, but brevity is no virtue at the cost of accuracy.

Phrasing the question

Having decided to include a 'vox pop' in the programme the producer, or the reporter working to him, must decide carefully the exact form of words to be used. The question is going to be addressed to someone with little preparation or 'warm-up' and so must be relatively simple and unambiguous. Since the object is to obtain opinions rather than a succession of 'yes/no' answers, the question form must be carefully constructed. Once decided, the same question is put each time otherwise the answers cannot sensibly be edited together. A useful question form in this context is: 'What do you think of . . .?' This will elicit an opinion which can if necessary be followed with the interviewer asking 'why' – this supplementary to disappear in the editing.

An example

What do you think of the proposal to raise the school-leaving age?

> It sounds all right but who's going to pay for it?
> I think it's a good idea, it'll keep the youngsters out of mischief.
> It'll not do *me* any good, will it?
> Bad in the short term, good in the long.
> (Why?)
> (Well) it'll cause an enormous upheaval over teachers' jobs and classrooms and things like that, but it's bound to raise standards overall eventually.
> I've not heard anything about it.
> The cost! – and that means higher taxes all round.
> I don't think it'll make much difference.
> (Why?)
> Because for those children who want to leave and get a job it'll be a waste and the brighter ones would have stayed on anyway.
> I think it's a load of rubbish, there's too much education and not enough work.

It is important that the question is phrased so that it contains the point to which reaction is required. In this example reference is made to the proposal to raise the school-leaving age to which people can respond even if they had not heard about it. This is much better than asking 'What do you think of the government's new education policy?'

In addition to testing opinion, the 'vox pop' can be used to canvass actual suggestions or collect facts, but where the initial response is likely to be short, a follow-up is essential. This question can be subsequently edited out. For example: Who is your favourite TV personality? The answer is followed by asking: Why? Another example: How often do you go to the cinema? – and then: Is this less often than you used to go? – or: Why would you say this is?

It is undoubtedly true that the more complex and varied the questioning, the more difficult will be the subsequent editing. The 'vox pop' producer must remember that he is not conducting an opinion poll or assembling data, he is making interesting radio which has to make sense in its limited context. The second example here may be useful in allowing the listener to compare his frequency of cinema-going with that of other people. But the producer must ask himself whether his intention would be equally well met, and more simply obtained, by the question 'What do you think of the cinema these days?'

A characteristic of the 'vox pop' is that in the final result, the interviewer's voice does not appear. The replies must be such that they can be joined together without further explanation to the listener and hence the technique is distinguished from simply a succession of interviews. The conversations should not be so complex that the interviewee's contribution cannot stand on its own.

Choosing the site

If the questioning is to be carried out among a specific group, this may itself dictate the place – on the docks, children leaving school, at the airport, etc. If the material is to be gathered generally, the site or sites chosen will be limited by technical factors, so as to permit easy editing at a later stage. These are to do with a reasonably low but essentially constant background noise level.

The listener expects to hear some background actuality and it would be undesirable to exclude it altogether. However, in essence, the broadcast is to consist of snatches of conversation in the form of remarks made off-the-cuff in a public place, and under these conditions immediate intelligibility is more difficult to obtain than in the studio. A side street will be quieter than a main road but a constant traffic background is preferable to intermittent noise. For this reason the interviewing site should not be near a bus stop, traffic lights or other road junction – the editing process becomes intrusively obvious if buses are made to disappear into thin air and lorries arrive from nowhere. Similarly, the site should be free from any sound which has a pattern of its own, such as music, public address announcements, or a chiming clock. Editing the speech so that it makes sense will be difficult enough without having to consider the effect of chopping up the background. A traffic-free pedestrian precinct or shopping arcade is often suitable, but a producer should avoid always returning to the same place – one of the attractions of the 'vox pop', in the general form as well as in the individual item, is its variety.

The recorder

The machine and its microphone are tested before leaving base, and on site a further check made to ensure an adequate speech level against the background noise. Some ten seconds of general atmosphere should be recorded to provide spare background for the fades in and out. From here on, the recording level control should not be altered otherwise the level of the background noise will vary. In order to maintain the same background level it is preferable to use a machine with a manual rather than an automatic recording control. Any AGC should therefore be switched out. Different speech volumes are compensated for by the positioning of the microphone relative to the speaker, and of course the normal working distance will be considerably less than in a studio.

To simplify the editing, only the actual replies need be recorded, and therefore a machine with a rapid and unobtrusive means of starting is a considerable asset – some types may be kept running but held on 'pause' until required; others may be started by a switch on the microphone.

Putting the question

It is normal for the novice reporter to feel shy about his or her first 'vox pop' but cases of assault on broadcasters are relatively rare. It may be helpful to remember that the passer-by is being asked to enter the situation without the benefit of any prior knowledge and is probably far more nervous. However, the initiative lies with the interviewer and he needs to adopt a positive technique. He should explain quickly who he is and what he wants, put the question, and record the reaction.

First, the reporter should be obvious rather than secretive. He stands in the middle of the pavement with his machine over his shoulder holding the microphone for all to see. It is most useful for the microphone to carry an identifying badge so that the approaching pedestrian can already guess at the situation, and if necessary take avoiding action. No one should be, or for that matter can be, interviewed against their will. Any potential but unwilling contributors should not be pursued or in any way harassed. In this sense, although the interviewer may receive the occasional rebuff, the contributors are only those who agree to stop and talk.

Seeing a prospective interviewee, the reporter approaches and says pleasantly 'Good morning. I'm from Radio XYZ'. At this the passer-by will either continue, protesting at how busy he is, or he will stop, being reassured by the truth of the statement since it confirms the station identification badge. He may possibly also be interested at the prospect of being on the radio. The reporter continues 'Can I ask you what you think of the proposal to keep all traffic out of the city centre' at which point he moves the microphone to within a foot or so of the contributor and switches on the recorder. In the chapter on interviewing, questions which began with 'Can I ask you' or 'Could you tell me' were generally disallowed on the grounds that they were superfluous; permission for the interview having already been granted, and that being unnecessary they were a waste of the listener's time. In the context of the 'vox pop' such a preamble is acceptable since it allows someone the courtesy of non-cooperation, and in any case the phrase will disappear in the editing.

The normal reaction of the 'man in the street' will vary from total ignorance of the subject, through embarrassed laughter and a collecting of thoughts, to a detailed or impassioned reply from someone who knows the subject well. All of this can be useful but there is likely to be a wastage rate of at least 50 per cent. If about ten replies are to be used, then twenty should be recorded. If the final result is to be around two minutes – and a 'vox pop' would seldom if ever exceed this – a total of four or five minutes of response should be recorded, by which time the interviewer will hope to have a diversity of views and some well-made argument.

Occasionally a group of people will gather round and begin a discussion. This may be useful, although inevitably some of it will be 'off-mic'. A developing conversation will be more difficult to edit and the 'one at a time' approach is to be preferred, although particularly with children more revealing comment is often obtained if they are within a group talking among themselves.

Whatever the individual response, the interviewer remains friendly and courteous. He will obviously want to give a good impression of the radio station he represents and will avoid becoming sidetracked into a discussion of the subject itself, station policy or last night's programme. He thanks each contributor for taking the trouble to stop and talk, remembering that it is they who have done him a favour.

The editing

Spontaneity, variety, insight and humour – these are the hallmarks of the good 'vox pop'. Listening to the material back at the base studio, the first step is to remove anything which is not totally intelligible. This must be done immediately, before the editor's ear becomes attuned to the sound. It is a great temptation to include a prize remark however imperfectly recorded on the basis of its being intelligible after a few playings under studio conditions! The rejection of material which is not of first-class technical quality is the first prerequisite to preventing the finished result from becoming a confusing jumble. Using a minidisc recorder allows the facility of rearranging the recorded tracks without the use of any other equipment. It is therefore possible to do some preliminary editing before getting back to the base studio. This requires care, however, since it is easy, in a hurry, to erase wanted material. In fact at this stage it is best not to erase anything, but simply to pull to the front of the disc the usable responses. Back at base, the material is downloaded to a computer and given its final edit. Recording on cassette requires dubbing before editing – onto either a hard disk or $\frac{1}{4}$-inch tape. Recording on reel-to-reel $\frac{1}{4}$-inch tape will need editing by physical cutting, although copying to a second tape will often be a quicker method of removing unwanted material and rearranging the remainder into the desired order.

The first piece of the finished vox pop needs to be a straightforward, clearly understood response to the question that will appear in the introductory cue material. The subsequent comments are placed to contrast with each other, either in the opinions expressed or their style. Men's voices will alternate with those of women, the young with the old, the local accent with the 'foreign', the 'pros' with the 'antis'. The interviewer's voice is not used, except that occasionally it may be useful to be reminded of the question half-way through. What must be avoided of course is its continual repetition. Sometimes the answers themselves are similar in which case sufficient should be used to indicate a consensus but not to become boringly repetitious. A problem can arise over the well-argued but lengthy reply which would be likely to distort the shape of the vox pop if used in its entirety. A permissible technique here is to cut it into two or three sections placing them separately within the final piece.

The editor needs a good comment to end on. Its nature will depend on the subject but it might be a view forcefully expressed, a humorous remark, or the kind of plain truth which often comes from a child. Good closing comments are not difficult to identify and the interviewer, when he has finished collecting his material, generally knows how his vox pop will end. The spare background noise is used as required to separate replies, with a few seconds at the beginning and at the end as a 'fade up' and 'fade down' under speech – so much better than 'banging it in' and 'chopping it off' on the air.

It should go without saying that the finished vox pop will broadly reflect the public response found by the interviewer. It is possible of course for the editing to remove all views of a particular kind so giving the impression that they do not exist. It may be that a producer would set out with the deliberate intention of demonstrating the overwhelming popularity of certain public attitudes – presumably those which accord with his own. Such manipulation, apart from betraying the trust which hopefully the listener has in him, is ultimately self-defeating. The listener does his own vox pops every day of his life – he will know whether or not the radio station is biased in its reflection of public opinion. Probably more than the broadcaster, the listener knows his own reality when he hears it.

Used properly the vox pop represents another colour in the broadcaster's palette. It provides contrast with studio material and in reflecting accurately what people are saying, it helps the listener to identify with the station and so enhances its credibility.

13

Listeners' letters

Radio is not a good medium by itself for establishing a genuine two-way contact. Listeners may feel that the broadcaster comes into their home and they may even get the impression that they know an individual presenter. However, this is at best a substitute companionship rather than a genuine pesonal interaction. The broadcaster/listener relationship – or perhaps that of the station/listener – can be made more real through the broadcaster's ability to reply to listener correspondence. It is a matter of station policy whether or not an individual presenter is encouraged to become involved in writing back to listeners. It is a time-consuming business and a hard-pressed station may not have the resources to do this. On the other hand, if a programme offers help, particularly for individual personal needs, then as with a phone-in, it must clearly honour its promise by meeting such requests which may also arrive by fax or e-mail.

Programme correspondence incoming to the station may be classified under two general headings – letters intended for use on-air, and those requiring a response only by mail.

On-air use

This category includes complaints, music requests and dedications, competition replies or letters written directly to a programme 'letter spot'; anything clearly intended for publication, i.e. to be read on-air. In general they either offer advice to the audience at large or ask for help with a personal problem. Letters like these can be a useful resource for programme makers – they frequently raise questions of interest and make comments of substance, and can easily provide the sole content for a specialist programme. They may be dealt with in a variety of ways:

1 Read by the presenter, who then responds.
2 Read by male/female readers, the programme presenter then responding.
3 Read by the presenter who then interviews an expert or introduces a discussion on the subject.

4   Letters on similar or related topics grouped together for subsequent response.
5   A letter dramatised as a sketch to illustrate the point being made, followed by a response by the presenter, an interviewee, or by a discussion.
6   Letter read by the presenter who without replying opens the discussion to the audience, inviting listeners to respond.

The producer of a letter spot may consider the following in arriving at the most appropriate format:

• To maximise listener involvement several letters should be dealt with in a single spot.
• A long letter may not be read in its entirety but extracts used to reflect accurately what the writer is saying.
• A letter with many questions should not monopolise a spot but could be used in parts, perhaps over several spots.
• To give variety of pace and vocal interest a spot may use more than one of the response forms listed above.
• Avoid reading out full addresses on the air or giving any information likely to endanger the writer (see pp. 178–9).

It is important that programme presenters do not read letters or reply to them in a patronising way, but respond to them as to a valued friend. Broadcasters are neither omniscient nor infallible and should always distinguish between a factual answer, best advice, and a personal opinion. Some basic and essential reference sources – books, databases, available experts, etc. – will provide answers for questions like, why does the leaning tower of Pisa lean? Or, how do you remove bloodstains from cloth? But it is a quite different matter replying to questions about how to invest your money, a cure for cataract, the causes of a rising crime rate, or why a loving God allows suffering. Nevertheless, without being glib or superficial it is possible for such a letter to spark off an interesting and useful discussion, or for the presenter to give an informed and thought-out response with:

   'On balance I would say that . . .'
or 'If you are asking for my own opinion I'd say . . . but let me know what you think.'

Research may have thrown up a suitable quote or piece of writing on the subject, which should of course always be attributed. Other phrases useful here are:

   'Experts seem to agree that . . .'
   'No one really knows, but . . .'

People write to radio stations either to get an issue aired publicly, or to get an authoritative reply. A letter is not simply programme fodder, but deserves the same level of consideration which its sender gave it. This becomes increasingly important for short-wave or long-distance broadcasters.

Off-air correspondence

Mail items not intended for broadcast include requests for listener verification or QSL cards, scripts, programme information and offers, merchandise purchasing, and programme follow-up ranging from specific advice to pastoral and personal counselling. Much of this can be dealt with by means of standard replies – perhaps with some details to be filled in – or by fact-sheets, information booklets, and so on. It is the time-consuming one-off reply which poses the genuine problem.

Complaints in particular need special attention. It is tempting for a busy station to disregard them, and it has to be said that many are likely to seem unreasonable resulting from an extreme or limited point of view. Nevertheless, if broadcasters are not to appear careless over alleged error and off-hand with the listener, such letters deserve a prompt but considered response at the appropriate level – programme presenter or producer, programme management or senior management – and if found to be of substance, a correction made. If the matter is serious and the station is genuinely in the wrong, an apology should be broadcast in the same slot as the offending transmission. The listening public deserves the highest standard of communication and, done well, a correction by letter or on-air can enhance a programme's reputation not only for the truth, but in its respect for the audience. See also pp. 286–7.

A larger category, we hope, comprises those letters which query or seek clarification over something that has been said – or which wish to debate the issue further. This poses another question – is the broadcaster responsible for questions which a programme may have raised in a listener's mind? A small station may have to ignore such correspondence – it represents a drain on resources without providing any airtime. Besides, a presenter may claim that the view of the listener even after a provocatively contentious programme is none of his business and that once the programme is over, the matter is finished. Furthermore, a philosophical, political or theological question raised may be outside his competence to answer.

On the other hand, a station may actively seek to develop the relationship with its listeners through follow-up. This is often best done not by the programme presenters or producer individually but by a separate group – either a specialist department of the station staff, or volunteers closely associated with the programme. Such broadcasting support services are extremely useful for educational and religious programmes in particular, where individual listeners may need direct personal help – perhaps struggling with learning a language, coming to terms with bereavement or coping with unemployment. Members of the replying group will need to be sensitive to any cultural differences which may exist between themselves and their correspondents. They will be selected as people experienced in a specialist field, but they will also rapidly develop their own expertise in this special form of one-to-one 'distance learning'. A computerised database enables the station to keep track of the correspondence and perhaps to anticipate further enquiries.

Another method is to put the listener in touch with a suitable college tutor, library, church, self-help group or other agency in his own vicinity. Question follow-up can then be pursued on a personal basis.

These are ways in which a station can extend its obligations as a public servant which radio communication alone cannot do. But it should not be undertaken lightly for it can be expensive in time, money and effort. It therefore requires the backing of a management policy which understands and values this additional form of listener contact.

14

Music programming

The filling of programme hours with recorded music is a universal characteristic of radio stations around the world. This is hardly surprising in view of the advantages discs have for the broadcaster. They represent a readily available and inexhaustible supply of high-quality material of enormous variety that is relatively inexpensive, easy to use and enjoyable to listen to. Before looking in detail at some of the possible formats and what makes for a successful programme, there are three important preliminaries to consider.

First, the matter of music copyright. Virtually every CD, cassette, and record label carries the words, 'all rights of the producer and of the owner of the recorded work reserved. Unauthorised public performance, broadcasting and copying of this record prohibited.' This is to protect the separate rights of the composer, publisher, performers and the recording company, who together enabled the disc to be made. The statement is generally backed by law – in Britain the Copyright, Designs and Patents Act (1988). It would obviously be unfair on the original artists if there were no legal sanctions against the copying of their work by someone else and its subsequent remarketing on another label. Similarly, broadcasters who in part earn their living through the effort of recording artists and others, must ensure that the proper payments are made regarding the use of their work. As part of a 'blanket' agreement giving 'authorised broadcasting use', most radio stations are required to make some form of return to the societies representing the music publishers and record manufacturers, indicating what has been played. In Britain these are, respectively, the Performing Right Society (PRS) and Phonographic Performance Ltd (PPL). It is the producer's responsibility to see that any such system is carefully followed.

Second, it must be said that in using records, broadcasters are apt to forget their obligation to 'live' music. Whatever the constraints on the individual station, some attempt should be made to encourage performers by providing opportunities for them to broadcast. Many recording artists owe their early encouragement to radio, and broadcasting must regard itself as part of the process which enables the first class to emerge.

Having reached the top, performers should be given the fresh challenge which radio brings.

Third, top-flight material deserves the best handling. It is easy to regard a record simply as a 'thing' instead of a person. On the air someone's reputation may be at stake. Basic operational technique must be faultless – levels, accurate talk-overs, fades, etc. Most important is that music should be handled with respect to its phrasing.

Attitudes to music

Music, like speech, comes in sentences and paragraphs. It would be nonsensical to finish a voice piece other than at the end of a sentence, and similarly it is wrong to fade a piece of music arbitrarily. A great deal of work has gone into its production and it should not be treated like water out of a tap, to be turned off and on at will – not unless the broadcaster is prepared to accept the degradation of music into simply a plastic filler material. The good operator therefore will develop an 'ear' for fade points. The 'talk-over' – an accurately timed announcement which exactly fits the non-vocal introduction of a song – provides a satisfying example of paying attention to such detail. Music handled *with respect to its phrasing* provides listening pleasure for everyone.

The presenter must accept the responsibility when music and speech are mixed through an automatic voice-over unit or 'ducker' so that whenever he speaks, the music is hurled into the background. It has its uses in particular types of high-speed DJ programme, but to use music as a semi-fluid sealant universally applied seems to imply that the programme has cracks which have to be frantically filled!

The broadcaster's attitude to music is often typified by his care in the treatment of CDs, cassettes and records. They are worth looking after. This includes an up-to-date library cataloguing system and proper arrangements for withdrawal and return; thus avoiding their being left lying around in the studio or production offices. Very many broadcasters who play music professionally have domestic stereo equipment over which they take meticulous care – they would not dream of touching the playing surface of a disc with their fingers.

Recordable CDs can store a wealth of material on their 99 possible tracks, from commercials and station jingles to individual presenter idents and programme inserts. Compilation CDs can be made to collect together the currently favoured music tracks, as well as being an excellent medium for station archives.

The programme areas now discussed in detail are: music formats, requests and dedications, guest programmes, and the DJ show.

Clock format

Designing a music programme on a 1-hour clockface has several advantages. It enables the producer/presenter to see the balace of the show between music and speech, types of music, and the spread of commercials; it is a great help in maintaining consistency when another presenter has to take over; and it enables format changes to be made with the minimum of disruption. It is a suitable method to use regardless of the length of the programme. The clock provides a solid framework from which a presenter may depart if need be, and pick up again just as easily. It imposes a discipline but allows freedom.

Starting with the audience, the producer begins by asking questions. Who do I wish to attract? Is my programme to be for a particular age or demographic group? This will obviously affect the music chosen, and a reasonable rule of thumb assumes that the musical taste of many people was formed in their teens. For example, listeners in their forties are likely to appreciate the hits of 25 years ago. Of course there are many categories of music which have their own specialist format or may be used to contribute to a programme of wider appeal. Beware, however, of creating too wide a contrast – the result is likely to please no one. The basic categories, which contain their own sub-divisions, can be listed as follows:

- Top 40
- Progressive rock
- Black soul/funk
- Rhythm and blues
- Disco-beat
- Adult contemporary
- Classic hits (the last 25 years)
- Golden oldies
- Jazz
- Folk
- Country
- Latin American
- Middle of the road
- Light classical, orchestral – operetta
- Classical, symphonic – opera, etc.

Having established the broad category there are infinite combinations of tempo and sound with which to achieve the essential variety within any chosen consistency:

- sound – rock band, big/small, heavy metal, synthetic
- tempo – slow, medium, med/bright, up-beat
- vocal – male, female, duo, group, newcomer, star
- era – twenties, thirties, forties, fifties, sixties . . . etc.
- ensemble – big band, string, military band, brass, orchestral, choral.

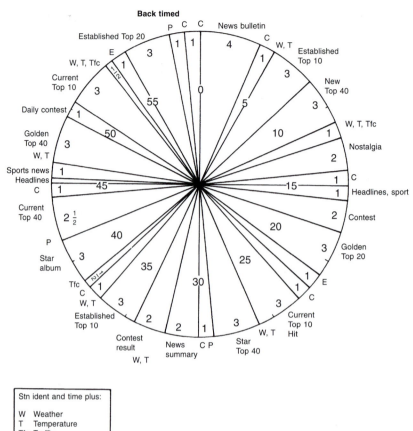

Figure 14.1   A clock format illustrating the concept of a breakfast time programme on the basis of:

- News on the hour, summary on the half-hour, headlines at quarters.
- Eight minutes of advertising in the hour, employing news-adjacency.
- Music/speech ratio of about 60:40.
- Generally MOR sound using current hits, 'golden' or yesteryear top 40, star name, and nostalgia tracks.
- Speech includes current information, sports news, 'short' programme contest and station daily competition.

Lengthy formatted programmes need not be constructed to have a strong beginning and end; the skill lies in the presenter providing the listener with a satisfying programme over whatever time-period the show is heard. The breakfast show and drive-time programmes are likely to be heard by many people over a relatively short time-span, perhaps 20–30 minutes. The afternoon output will probably be heard in longer durations. The producer sets out to meet the needs of all his audience – so the two key questions are: Does the plan contain the essential elements in the estimated normal period of listening? Does it contain the necessary variety over a longer time-span?

Station practice indicates the following guidelines:

- Come out of news bulletins on a non-vocal up-tempo sound. You need to restore of music pace and avoid any unfortunate juxtaposition of news story and lyric.
- When programming new or unfamiliar music, place solid hits or known material on either side of it – 'hammocking'.
- Spread items of different types evenly up to the hour as well as on the downside.
- Mix items of different durations, avoid more than two items of similar length together.
- Occasionally break up the speech – music – speech sandwich by running discs back to back, i.e. use the segue or segue plus talk-over.
- Place regular items at regular times.

Computerised selection

Many stations regulate the use of music in a totally systematic way, for example to prevent the same tracks being used in adjacent programmes. Entering titles as data, a computer is programmed to provide the right mix of material with music styles, performers, composers, etc. appearing at the desired frequency. The scheduling is then in the hands of the computer programmer who decides the rotation of items, how many times a track will be played in a given period, and what type and length of music is appropriate for different times of day. The computer keeps a history of the music played which not only provides the official music returns, but is the source for answering listener enquiries.

A computerised system is used to generate a weekly playlist which is mandatory for presenters who are given no choice over their music. Less restricting is the playguide. This requires presenters to select a given percentage of their programme from material determined by the programme controller; the remainder they choose themselves. Station policy invariably lays down the extent of individual freedom in the matter of music choice.

Requests and dedications

In presenting a request show it is all too easy to forget that the purpose is still to do with *broad*casting. There is a temptation to think only of those who have written in, rather than of the audience at large. Until the basic approach is clarified, it is impossible to answer the practical questions which face the producer. For example, to what extent should the same

track be played in successive programmes because someone has asked
for it again? Is there sufficient justification in reading out a lengthy list
of names simply because this is what appears on the request card? While
the basis of the programme is clearly dependent on the initiative of the
individual listener who requests a record, the broadcaster has a responsibility
to all listeners, not least the great majority who do not write. The programme
aims may be summarised:

1   To entertain the general audience.
2   To give especial pleasure to those who have taken the trouble to send
    a request.
3   To foster goodwill by public involvement.

Programmes are given a good deal of individual character by the presenter
who states his own guidelines. He may deal only in requests related to
birthdays, weddings or anniversaries, or he may insist that each request
is accompanied by a personal anecdote, joke or reminiscence to do with
the music requested. References to other people's nostalgia can certainly
add to the general entertainment value of the programme. The presenter
should be consistent about this and it is important that style does not take
over from content. The declared intention of the programme is to play
music and it is therefore this rather than speech which remains the central
ingredient.

A further variation is to make a point of including dedications as well
as requests, i.e. cards not related to a specific piece of music but which
can be associated with any item already included in the programme. By
encouraging 'open' cards of this type, the programme can carry more
names and listeners therefore have a greater chance of hearing themselves
mentioned.

Having decided the character and format of the programme there are
essential elements in its preparation.

Choosing music

With music items of $2\frac{1}{2}-4$ minutes each, there will be some eight or nine
tracks in each half hour of non-advertising air time. This allows for about
a minute of introduction, signature tunes, etc. Given the volume of letters
and cards received, it soon becomes clear to what extent the actual requests
can be met. The proportion which can be dealt with may be quite small
and assuming that their number exceeds the capacity to play them, a
selection process is necessary.

The criteria of selection will include the presenter's desire to offer an
attractive programme overall, with a variety of music consistent with the
programme policy. It may be limited to the current Top 40, to pop music
generally, or it may specifically deal with one area, for example gospel

music. On the other hand, the choice may be much wider to include popular standards, light classical music or excerpts from symphonic works. The potential requester should know what kind of music a particular programme offers.

Another principle of selection is to choose requests that are likely to suit the presenter's remarks. Those which will make for an interesting introduction, an important or unusual event, an amusing remark or a particularly topical reference. It is frequently possible to combine requests using comments from a number of cards in order to introduce a single piece of music. The danger here is to become involved in several lists of names which may delight those who like hearing themselves referred to on the air, but which can become boring to the general listener. It is, however, a useful method of including a name check while avoiding a particular choice of music either because to do so would be repetitious, or because it is out of keeping with the programme – or of course because the station does not have the named track.

Item order

After selecting the music, a decision has to be made about its sequence. This should not be a matter of chance for there are positive guidelines in building an attractive programme. 'Start bright, finish strong' is an old music hall maxim and it applies here. A tuneful or rhythmic familiar up-tempo 'opener' with only a brief speech introduction will provide a good start with which the general listener can identify. A slower piece should follow and thereafter the music can be contrasted in a number of ways: vocal/non-vocal, female vocal/male vocal, group vocal/solo vocal, instrumental/orchestral, slow/fast, familiar/unfamiliar, and so on. It is sensible to scatter the very popular items throughout the programme and to limit the material which may be entirely new. Careful placing is required with slow numbers which need to be followed by something brighter. The order in which music is played of course affects all music programmes whether or not they are based on the use of records.

Prefading to time

The last piece of music can be chosen to provide 'the big finish' and a suitable finale will often be a non-vocal item. This allows it to be faded-in under the presenter's introduction. It may have been timed to end a little before the close of the programme, so leaving room for the final announcements and signature tune. 'Prefading to time' (not to be confused with 'prefade' – audition or pre-hear) or 'back timing' is the most common device for ensuring that record programmes run to time. A closing signature tune will almost certainly be prefaded to time.

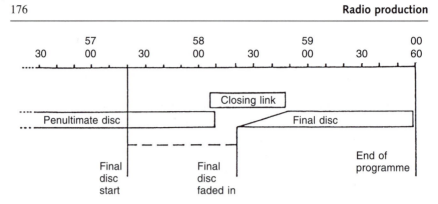

Figure 14.2 Prefading to time. This is the most usual method of ensuring that programmes run to time. In this example, the final disc is 2′40″ long. It is therefore started at 57′20″ but not faded up until it is wanted as the closing link finishes. It runs for a further 1′10″ and brings the programme to an end on the hour

In a similar way items within a programme, particularly a long one, can be subject to this technique to provide fixed points and to prevent the overall timing drifting.

Preparing letters and cards

Presenters using letters from listeners will know that they are not always legible and cannot be used in the studio without some form of preparation. Some type out the basic information to ensure clarity, others prefer to work directly from the original cards and letters. It is important that names and addresses are legible so that the presenter is not constantly stumbling, or sounding as though the problem of deciphering his correspondent's handwriting is virtually insurmountable. In reality, this may often be the case, but a little preparation will avoid needlessly offending the people who have taken the trouble to write and on whom the programme depends. In this respect, particular care has to be taken over personal information such as names and ages, and it goes without saying that such things should be correctly pronounced. This also applies to streets, districts, hospitals, wards, schools and churches where an incorrect pronunciation – even though it be caused by illegible handwriting – immediately labels the presenter as a 'foreigner' and so hinders listener identification. Such mispronunciations should be strenuously avoided, particularly for the community broadcasting station, and prior reference to telephone directories, maps, and other local guides is essential.

Reading through the cards beforehand enables the presenter to spot dates, such as anniversaries, which will be past by the time of the broadcast. This can then be the subject of a suitable apology coupled with a message of goodwill, or alternatively the request can be omitted. What should not happen is for the presenter to realise on-air during his introduction that the event which the writer anticipates has already happened. Not only does this sound unprofessional, but it gives the impression that the presenter does not really know what he is talking about, and worse, does not care. Anything which hinders the rapport between listener and presenter will detract from the programme.

In order to accommodate more requests and dedications, cards can be grouped together, sharing a single piece of music. The music need not necessarily be precisely what was asked for, providing that the presenter makes it quite clear what he is doing. The piece should, however, be of a similar type to that requested, for example by the same artist.

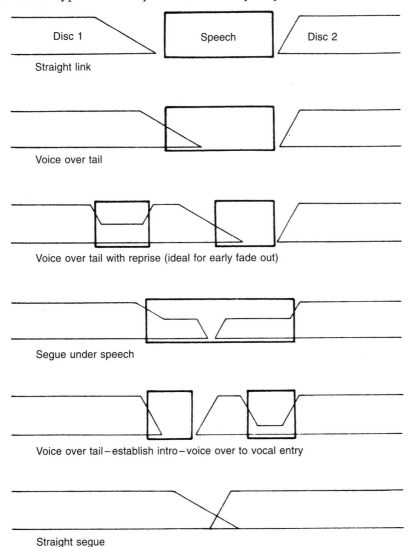

Figure 14.3 Six ways of going from one music item to the next, with or without a speech link. The use of several different methods helps to maintain programme variety and interest

Such preparation of the spoken material makes an important contribution to the presenter's familiarity with the programme content. While it may be his intention to appear to be speaking informally 'off the cuff', matters such as pronunciation, accuracy of information, relevance of content, and timing need to be worked on in advance. The art is to do all this and still retain a fresh, 'live', ad-lib sound. However, the experienced presenter knows that the best spontaneity contains an element of planning.

Programme technique

On the air, each presenter must find his or her own style and be true to it. No one person or programme will appeal to everyone but a loyal following can be built up through maintaining a consistent approach. A number of general techniques are worth mentioning.

Never talk over vocals. Given adequate preparation, it is often pleasing to make an accurate talk-over of the non-vocal introduction to a record. But talking over the singer's voice can be muddling, and to the listener may sound little different from interference from another station.

Avoid implied *criticism of the listener's choice* of music. The tracks may not coincide with your taste but they represent the broadcaster's intention to encourage public involvement. If a request is unsuitable, it is best left out.

Do not play less than one minute of anyone's request, or the sender will feel cheated. If the programme timing goes adrift it is generally better to drop a whole item than to compress. If the last item, or the closing signature tune, or both are 'prefaded to time' the programme duration will look after itself.

If by grouping requests together, there are several names and addresses for a single record, prevent the information from *sounding like a roll call*. Break up the list and intersperse the names with other remarks.

Develop the habit of talking *alternately* to the *general listener* and to the *individual listener* who asked for the record. For example:

> 'Now here's a card from someone about to celebrate 50 years of married bliss – that's what it says here and she's Mrs Jane Smith of Highfield Road, Mapperley. Congratulations Mrs Smith on your half century, you'd like to dedicate this record with all your love to your husband John. An example to the rest of us, – congratulations to you too, John. The music is from one of the most successful shows of the sixties . . .'

Avoid remarks which combined with the address may pose any kind of *risk to the individual*.

> '. . . I'm asked not to play it before six o'clock because there's nobody at home till then . . .'

'. . . she says she's living on her own but doesn't hear too well . . .'
'. . . please play the record before Sunday the 18th because we'll all be away on holiday after that . . .'
'. . . he says his favourite hobby is collecting rare stamps.'

To reduce such risks, some presenters omit the house detail, referring only to the street and town. However, house numbers can generally be discovered by reference to voters' lists and other directories. Broadcasters must always be aware of the possible illegal use of personal information.

It is often a wise precaution to keep the cards and letters used in a programme for at least a week afterwards in order to deal with music enquiries or other follow-up which may be required.

Guest programmes

Here, the regular presenter invites a well-known personality to the programme and plays his or her choice of music. The attraction of hearing artists and performers talk about music is obvious but it is also of considerable interest to have the lives of others, such as politicians, sportsmen and businessmen, illuminated in this way.

The production decisions generally centre on the ratio of music to speech. Is the programme really an excuse for playing a wide range of records or primarily a discussion with the guest but with musical punctuation? Certainly it is easy to irritate the listener by breaking off an interesting conversation for no better purpose than to have a musical interlude, but equally a tiny fragment only of a Beethoven symphony can be very unsatisfying. The presenter, through his combination of interview style and links into and out of the music, must ensure an overall cohesiveness to prevent the programme sounding 'bitty'. While it may be necessary to arrive at a roughly half and half formula, the answer could be to concentrate on the music where the guest has a real musical interest but to increase the speech content where this is not the case, using fewer but perhaps slightly longer music inserts.

Once again the resolution of such questions lies in the early identification of a programme aim appropriate to the target audience.

DJ programmes

The radio disc jockey defies detailed categorisation. His or her task is to be unique, to find and establish a distinctive formula different from all other DJs. The music content may vary little between two competing programmes and in order to create a preference the attraction must lie in the way it is presented. To be successful therefore, the DJ's personality and programme style must not only make contact with the individual listener, but in themselves be the essential reason for the listener's attention. The style may be elegant or earthy, raucous or restrained, but for any one presenter it should be consistent and the operational technique first class.

The same rules of item selection and order apply as those already identified. The music should be sufficiently varied and balanced within its own terms of reference to maintain interest and form an attractive whole. Even a tightly formatted programme, such as a Top 40 show, will yield in all sorts of ways to imaginative treatment. One of the secrets is to be absolutely certain of the target audience. Top 40 for the 20–30 age group is radically different from Adult Contemporary for 25–40-year-olds. Different again is Nostalgia for the 40+. The personal approach to this type of broadcasting differs widely and can be looked at under three broad headings.

The low-profile DJ

Here the music is paramount and the presenter has little to say. His job is to be unobtrusive. The purpose of the programme may be to provide background listening and all that is required is the occasional station identification or time check. In the case of a classical music programme the speech/music ratio should obviously be low. The listener is easily irritated by a presenter who tries to take over the show from Berlioz and Bach. The low-profile DJ has to be just as careful over what he says as his more voluble colleagues.

The specialist DJ

Experts in their own field of music can make excellent presenters. They spice their introduction with anecdotes about the artists and stories of happenings at recording sessions, as well as informed comment on performance comparisons and the music itself. Jazz, rock, opera and folk all lend themselves to this treatment. Often analytical in approach, the DJ's job is to bring alive the human interest inherent in all music. The listener should obviously enjoy the tracks but half the value of the programme is derived from hearing authoritative, possibly provocative, comment from someone who knows the field well.

The personality DJ

This is the most common of all DJ types. He must do more than just play tracks with some spontaneous ad-libs in between. However popular the music, this simple form of presentation soon palls. The DJ must communicate his own personality, creating a sense of friendship with his audience. He therefore never embarrasses his listener, either through incompetence or bad taste – he must entertain. To do this well, programme after programme, requires two kinds of preparation.

The first is in deciding what to say and when. This means listening to the music beforehand to decide the appropriate places for a response to the words of a song, a jokey remark or other comment, where to place a listener's letter, quiz question or phone call. The chat between records should be thought about in advance so that it does not sound pedestrian, becoming simply a repetitive patter. All broadcast talk needs some real substance containing interest and variety. This is not to rule out entirely the advantages of spontaneity and the 'fly by the seat of your pants' approach. The self-operating DJ, with or without a producer, is often capable of creating an entertaining programme, making it up as he goes along. Undoubtedly though, such a broadcaster is even better given some preparation time.

The programme may also contain identifications, weather and traffic information, commercials, time checks, trails for other programmes and news. It may contain so many speech items that it is better described as a sequence rather than a DJ show, and this is developed further in the next chapter. But no presenter should ever be at a loss as to what to do next. He must know in advance what he wants to say, and be constantly replenishing his stock of anecdotes. Where possible these should be drawn from his own observation of the daily scene. Certainly for the local radio DJ, the more he can develop a rapport with his own area, the more his listeners will identify with him. The preparation of the programme's speech content will also include the timing of accurate talkovers and any research.

When a DJ is criticised for talking too much, what is often meant is that he is not interesting enough, i.e. there are too many words for what he has to convey. It is possible to correct this by talking less, but similarly the criticism will disappear if the same amount of speech is used to carry less waffle and more substance. Much of what is said may be trivial, but it should still be significant for the listener through its relevance and point of connection. So talk about things to which the listener can relate. Develop that rapport by asking – what information, what entertainment, what companionship does my listener need? If you don't know, go and meet some of them. And remember that there is a key factor in establishing your credibility, it's called professional honesty.

The second kind of preparation for a DJ, and where appropriate his producer, is in actually making additional bits and pieces of programme material which will help to bring the show alive. Probably recorded for accurate cueing, these may consist of snatches of conversation, sound effects, funny voices on echo, chords of music, and so on. Presenters may even create extra 'people', playing the roles themselves on the air, talking 'live' to their own recordings. Such characters and voices can develop their own personalities appearing in successive programmes to become very much part of the show. Only the amount of time which is set aside for preparation and the DJ's own imagination sets limits on what can be achieved in this way.

For the most part such inserts are very brief but they enliven a DJ's normal speech material adding an element of unpredictability and increasing the programme's entertainment value.

Whether the programme is complex or simple, the personality DJ should, above all, be fun to listen to. But while the show may give the impression of a spontaneous happening, sustained success is seldom a matter of chance. It is more likely to be found in a carefully devised formula and a good deal of preparation and hard work.

15

Magazines and sequences

Of all programme types, it is the regular magazine, or lengthy sequence, which can so easily become boring or trivial by degenerating into a ragbag of items loosely strung together. To define the terms, a magazine is usually designed with a specific audience in mind, and tightly structured with the emphasis on content. A sequence is generally longer – between 1 and 4 hours – often daily, using music with a wide audience appeal, and with an emphasis on the presentation. For both, the major problem for the producer is how best to balance the need for consistency with that of variety. Clearly there has to be a recognisable structure to the programme – after all, this is probably why the listener switched on in the first place. An obvious policy of marketing, which applies to radio no less than to any other product, is to build a regular audience by creating positive listener expectations, and then to fulfil or, better still, exceed them. The most potent reason for tuning in to a particular programme is that the listener liked what he heard last time. This time, therefore, the programme must be of a similar mould, not too much must be changed. It is equally obvious, however, that the programme must be *new* in the sense that it must have fresh and updated content and contain the element of surprise. The programme becomes boring when its content is too predictable, yet it fails if its structure is obscure. It is not enough simply to offer the advice 'keep the format consistent but vary the content'. Certainly this is important but there must be consistencies too in the intellectual level and emotional appeal of the material. From edition to edition there must be the same overall sense of style.

Since we have so far borrowed a number of terms from the world of print, it might be useful to draw the analogy more closely.

The newspapers and magazines we buy are largely determined by how we reacted to the previous issue. To a large extent purchases are a matter of habit and although some are bought on impulse, or by accident, changes in readership occur relatively slowly. Having adopted our favourite periodical, we do not care to have it tampered with in an unconsidered way. We develop a personal interest in the typography, page layout, length of feature article or use of pictures. We know exactly where to find the sports page, crossword or favourite cartoon. We take a paper which appeals to us as individuals; there is an emotional link and we can feel distinctly annoyed should a new editor change the typeface, or start moving things around when, from the fact that we bought it, it was all right as it was. In other words the consistency of a perceived structure is important since it leads to a reassurance of being able to find your way around, of being able to use the medium fully. Add a familiar style of language, words that are neither too difficult nor too puerile, sentences which avoid both the pompous and the servile, captions which illuminate and not duplicate; and it is possible to create a trusting bond between the communicator and the reader, or in our case the listener. Different magazines will each decide their own style and market. It is possible for a single publisher, as it is for the manager of a radio station, to create an output with a total range aimed at the aggregate market of the individual products.

For the individual producer, his crucial decision is to set the emotional and intellectual 'width' of his programme and to recognise when he is in danger of straying outside it.

To maintain programme consistency then several factors must remain constant. A number of these are now considered.

Programme title

This is the obvious signpost and it should both trigger memories of the previous edition and provide a clue to content for the uninitiated. Titles such as 'Farm', 'Today', 'Sports Weekly', and 'Woman's Hour' are self-explanatory. 'Roundabout', 'Kaleidoscope', 'Miscellany', and 'Scrapbook' are less helpful except that they do indicate a programme containing a number of different but not necessarily related items. With a title like 'Contact', or 'Horizon', there is little information on content, and a subtitle is often used to describe the subject area. Sequences, like DJ programmes, are often known by the name of the presenter – 'The Jack Richards Show' – but a magazine title should stem directly from the programme aims and the extent to which the target audience is limited to a specialist group.

Signature tune

The long sequence is designed to be listened to over any part at random – to dip in and out of. A signature tune is largely irrelevant, except to distinguish it from the previous programme serving as an additional signpost intended to make the listener turn up the volume. It should also convey something of the style of the programme – lighthearted, urgent, serious, or in some way evocative of the content. Fifteen seconds of the right music can be a useful way of quickly establishing the mood. Magazine producers, however, should avoid the musical cliché. While 'Nature

Notebook' may require a pastoral introduction, the religious magazine will often make strenuous efforts not to use opening music that is too churchy. If the aim is to attract an audience which already identifies with institutionalised religion, some kind of church music may be fine. If, on the other hand, the idea is to reach an audience which is wider than the churchgoing or sympathetic group, it may be better to avoid too strong a church connotation at the outset. After all, the religious magazine is by no means the same as the Christian magazine or the church programme.

Transmission time

Many stations construct their daily schedule with a series of sequences in fixed blocks of 3 or 4 hours. It is obviously important to have the right presenter and the right style of material for each time slot. The same principle holds for the more specialist magazine. Regular programmes must be at regular times and regular items within programmes given the same predictable placing in each programme. This rule has to be applied even more rigorously as the specialisation of the programme increases.

For example, a half-hour farming magazine may contain a regular three-minute item on local market prices. The listener who is committed to this item will tune in especially to hear it even though he may not bother with the rest of the programme. A sequence having a wider brief, designed to appeal to the more general audience, is more likely to be on in the background and it is therefore possible to announce changes in timing. Even so, the listener at home in the mid-morning wants the serial instalment, item on current affairs, or recipe spot at the same time as yesterday since it helps to orientate the day.

The presenter

Perhaps the most important single factor in creating a consistent style, the presenter regulates the tone of the programme by his approach to the listener. He or she can be outgoing and friendly, quietly companionable, informal or briskly businesslike, or knowledgeable and authoritative. It is a consistent combination of characteristics, perhaps with two presenters, which allows the listener to build a relationship with the programme based on 'liking' and 'trusting'. Networks and stations which frequently change their presenters, or programmes which 'rotate' their anchor people are simply not giving themselves a chance. Occasionally you hear the justification of such practice as the need 'to prevent people from becoming stale', or worse, 'to be fair to everyone working on the programme'. Most programme directors would recommend a six-month period as the minimum for a presenter on a weekly programme, and three months for a daily show. Less than this and he may hardly have registered with the listener at all.

In selecting a presenter for a specialist magazine, the producer may find himself with a choice of either a good broadcaster or an expert in the subject. Obviously the ideal is to find both in the same person, or through training to turn one into the other – the easier course is often to enable the latter to become the former. If this is not possible, an alternative is to use both. Given a strict choice the person who knows his material is generally preferable. Credibility is a key factor in whether or not a specialist programme is listened to, and expert knowledge is the foundation, even though it may not be perfectly expressed. In other words, if we have a doctor for the medical programme and a gardener for the gardening programme, should there be children for the children's show, and a disabled person introducing a programme for the handicapped? In a magazine programme for the blind there may be a bit more paper shuffling 'off mic', and the Braille reading may not be as fluent as with a sighted reader, but the result is likely to have much more impact and be far more acceptable to the target audience.

Linking style

Having established the presenter, or presenters, and assuming he or she will write, or at least rewrite, much of the script, the linking material will have its own consistent style. The way in which items are introduced, the amount and type of humour used, the number of time checks, and the level at which the whole programme is pitched will remain constant. The links of course refer to the various items, as discussed under the heading of 'Cue material', but they also enable the presenter to give additional information, personalised comment or humour. The 'link-person' is much more than a reader of cue material – announcements between items should extend beyond the simple 'this is' and 'that was'. It is interesting to speculate on the function of mortar in the building of a house. Does it keep the bricks apart or hold them together? It does both of course, and so it is with the presenter. Through this handling of the links he separates and delineates while at the same time creating a cohesive sense of style.

Information content

The more local a sequence becomes, the more specific and practical can be the information it gives. It may be either carried in the form of regular spots at known times or simply included in the links. If a programme sets out with the intention of becoming known for its information content, the spots must be distinctive, yet standardised in terms of timing, duration, style, 'signposting', introductory ident or sound effect.

The types of useful information will naturally depend on the particular needs of the audience in the area covered by the station. The list is wide ranging and typical examples for inclusion in the daily programme are:

News reports	Time checks
Weather	River conditions (for anglers)
Traffic information	Mobile library services
Sports results	Tide times
Tonight's TV	Changes to ferry times
Late-night chemists	Sea state – coastal waters
Road works for motorists	Shipping movements
Pollen count	Lighting-up times
Pavement works for the blind	What's on – entertainment
Rail delays	Top Twenty chart
Airport information	Review of papers/journals
Local flying conditions	Club meetings
Shopping prices	Fatstock prices for farmers
Financial market trends	Blood donor sessions
Racing information	Station identification

Programme construction

The overall shape of the programme will remain reasonably constant. The proportion of music to speech should stay roughly the same between editions, and if the content normally comprises items of from three to five minutes' duration ending with a featurette of eight minutes, this structure should become the established pattern. This is not to say that a half-hour magazine could not spend fifteen minutes on a single item given sufficient explanation by the presenter. But it is worth pointing out that by giving the whole or most of a programme over to one subject, it ceases to be a magazine and instead becomes a documentary or feature. There is an argument to be made in the case of a specialist programme for occasionally suspending the magazine format and running a one-off 'special' instead, in which case this should of course be aimed at the same audience as the magazine it replaces. Another permissible variation in structure is where every item has a similar 'flavour' – as in a 'Christmas edition', or where a farming magazine is done entirely at an agricultural show.

Such exceptions are only possible where a standard practice has become established for it is only by having 'norms' that one is able to introduce the variety of departing from them.

Sequences are best designed in the clock format described in the previous chapter.

Programme variety

Each programme must create fresh interest and contain surprise. First, the subject matter of the individual items should itself be relevant and new to the listener. Second, the treatment and order of the items need to highlight the differences between them and maintain a lively approach to the listener's ear. It is easy for a daily magazine, particularly the news magazine, to become nothing more than a succession of interviews. Each may be good enough in its own right, but heard in the context of other similar material, the total effect may be worthy but dull. In any case, long stretches of speech especially by one voice should be avoided. Different voices, locations, actuality, and the use of music bridges and stings should produce an overall effect of brightness and variety – not necessarily superficial, but something people actually look forward to.

Programme ideas

Producing a good programme is one thing, sustaining it day after day or week after week, perhaps for years, is quite another. How can a producer, with little if any staff assistance, set about the task of finding the items necessary to keep the programme going? First, he is never off duty but is always wondering if anything he sees or hears will make an item. He records even the passing thought or brief impression, probably in a small notebook which he always carries. It is surprising how the act of writing down even a flimsy idea can help it to crystallise into something more substantial. Second, through a diary and other sources he has advance information on anniversaries and other future events. Third, he cultivates a wide range of contacts. This means that he reads, or at least scans, as much as he can; newspapers, children's comics, trade journals, parish magazines, the small ads, poster hoardings, the internet – anything which experience has shown can be a source of ideas.

He gets out of the studio and walks through his territory – easier in local radio than for a national broadcaster! He is a good listener, both to the media and in personal conversation. He is aware of people's problems and concerns, and of what makes them laugh. He encourages his contributors to come up with ideas and if he has little money to pay them, at least he makes sure they get the credit for something good. If you are too authoritarian or adopt a 'know it all' attitude, people will leave you alone. But by being available and open, people will come to you with ideas. Welcome correspondence. It is hard work but the magazine/sequence producer soon develops a flair for knowing which of a number of slender leads is likely to develop into something for his programme.

Having decided to include an item on a particular subject, the producer has several options on the treatment he can employ. A little imagination will prevent his programme from sounding 'samey' and he should consider the extent to which he can increase the variety of his programme by the use of the following radio forms.

Voice piece

A single voice giving information as with a news bulletin, situation report, or events diary. This form can also be used to provide eye-witness commentary, or tell a story of the 'I Was There' type of personal reminiscence.

A voice piece lacks the natural variety of the interview and must therefore have its own colour and vitality. In style it should be addressed directly to the listener – pictorial writing in the first person and colloquial delivery can make compelling listening. But more than this, there must also be a *reason* for broadcasting such an item. There needs to be some special relevance – a news 'peg' on which to hang the story. The most obvious one lies in its immediacy to current events. Topicality, trends or

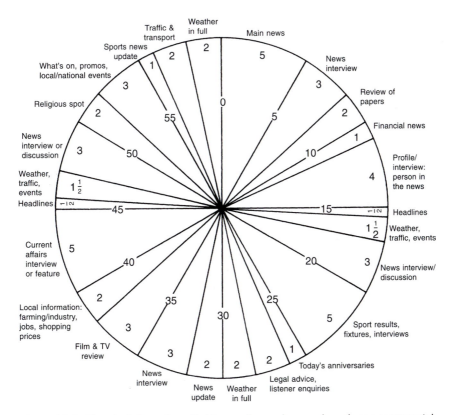

Figure 15.1 The clock format applied to an all-speech, news-based, non-commercial, breakfast sequence. Typically programmed from 6.30 am to 9.00 am, it requires a range of news, weather, sport and information sources, plus five interviews in the hour – live, telephoned or recorded. With repeats this might total nine in the $2\frac{1}{2}$ hours. Interviewees will be drawn mainly from politicians, people in the news, officials, 'experts', and media observers. Large organisations will have their own correspondents and local stringers. Presentation will include station idents, time, temperature, promos and incidental information

ideas relevant to a particular interest group, innovation and novelty, or a further item in a useful or enjoyable series are all factors to be used in promoting the value of items in a magazine.

Interview

The types of interview (see Chapter 3) include those which challenge reasons, discover facts, or explore emotions. This sub-heading also includes the vox pop or 'man in the street' opinion interview. A further variation is the question and answer type of conversation done with another broadcaster – sometimes referred to as 'Q & A' or a two-way. Using a specialist correspondent in this way the result is often less formal, though still authoritative.

Discussion

Exploration of a topic in this form within a magazine programme will generally be of the bi-directional type consisting of two people with opposing, or at least non-coincident, views. To attempt in a relatively brief item to present a range of views, as in the 'multi-facet' discussion, will often lead to a superficial and unsatisfactory result. If a subject is big enough or important enough to be dealt with in this way, the producer should ask himself whether it shouldn't have a special programme.

Music

An important ingredient in achieving variety, music can be used in a number of ways:

1 As the major component in a sequence.
2 An item, concert performance or record featured in its own right.
3 A new item reviewed.
4 Music which follows naturally upon the previous item. For example, an interview with a pianist about to make his concert debut followed by an illustration of his work.
5 Where there is a complete change of subject, music can act as a link – a brief music 'bridge' may be permissible. This is particularly useful to give 'thinking time' after a thoughtful or emotional item where a change of mood is required.

Music should be used as a positive asset to the programme and not merely to fill time between items. It can be used to supply humour or provide wry comment on the previous item, e.g. a song from *My Fair*

Lady could be a legitimate illustration after a discussion about speech training. Its use, however, should not be contrived merely because its title has a superficial relevance to the item. It would be wrong, for example, to follow a piece about an expedition to the Himalayas with 'Climb Every Mountain' from the *Sound of Music*. For someone who looks no further than the track notes there appears to be a connection, but the discontinuity of context would lead to accusations of poor judgement.

One of the most difficult production points in the use of the medium is the successful combination of speech and music. Music is much more divisive of the audience since listeners generally have positive musical likes and dislikes. It is also very easy to create the wrong associations, especially among older people, and real care has to be taken over its selection.

Sound effects

Like music, effects or actuality noises in a magazine can add enormously to what might otherwise be a succession of speech items. They stir the memory and paint pictures. An interview on the restoration of an old car would surely be accompanied by the sound of its engine and an item on dental techniques by the whistle of a high-speed drill. The scene for a discussion on education could be set by some actuality of playground or classroom activity, and a voice piece on road accident figures would catch the attention with the squeal of brakes. These things take time and effort to prepare and, if overdone, the programme suffers as it will from any other cliché. But used occasionally, appropriately and with imagination, the programme will be lifted from the mundane to the memorable. You do not have to be a drama producer to remember that one of the strengths of the medium is the vividness of impression which can be conveyed by simple sounds.

Listener participation

Daily sequences in particular like to stimulate a degree of audience involvement. Again, the producer has several ways of achieving this:

1 *Requests* for music or for a particular subject to be discussed, or asking for a favourite item to be repeated.
2 *Letters spot* which acts as a follow-up to the previous programmes, or a general correspondence column of the air.
3 *Competitions* are a good method of soliciting response. These could be in reply to quizzes and on-air games with prizes offered, or simply for the fun of it.
4 A *phone-in* spot in a live magazine helps the sense of immediacy and

can provide feedback on a particular item. Placed at the end, it can allow listeners the opportunity of saying what they thought of the programme as a whole. Used in conjunction with a quiz competition it obviously allows for answers to be given and a result declared within the programme.

5   A *'helpline'* is a useful item in a lengthy sequence – putting people with a specific need in touch with possible sources of help. For a small community station such links might vary from someone's search for a particular book or 'lost and found' information to a noticeboard of job vacancies. Such requests are normally broadcast without charge but the station generally takes no responsibility for the outcome, a listener responding to an enquiry being put in touch with the originator off-air.

6   *Casual visitors* to the radio station may be persuaded to become broadcasters for a short while and take part in a special spot, either live or by editing their recorded comments accumulated since the previous programme. Alternatively using a mobile facility such as a radio car, the programme can visit its listeners by dropping in on a home discussion group or factory meeting.

Listener participation elements need proper planning but part of their attraction lies in their live unpredictability. The confident producer or presenter will know when to stay with an unpremeditated turn of events and extend an item which is developing unexpectedly well. He or she will have also determined beforehand the most likely way of altering the running order to bring the programme back on schedule. In other words in live broadcasting the unexpected always needs to be considered.

Features

A magazine will frequently include a place for a package of material dealing with a subject in greater depth than might be possible in a single interview. Often referred to as a featurette, the general form is either person centred – 'our guest this week is . . .'; or place centred – 'this week we visit . . .'; or topic centred – 'this week our subject is . . .'.

Even the topical news magazine, in which variety of item treatment is particularly difficult to apply, will be able to consider putting together a featurette comprising interview, voice piece, archive material, actuality and links, even possibly music. For instance, a report on the scrapping of an old ferry boat could be run together with the reminiscences of its former skipper, describing some perilous exploit, with the appropriate sound effects in the background. Crossfade to breaker's yard and the sound of lapping water. This would be far more interesting than a straight report.

The featurette is a good means of distilling a complex subject and presenting its essential components. The honest reporter will take the crux of an argument, possibly from different recorded interviews, and present them in the context of his own links. They should then form a logical, accurate and understandable picture on which the listener can base an opinion.

Drama

The weekly or daily serial or book reading has an established place in many programmes. It displays several characteristics which the producer is attempting to embody – the same placing and introductory music, a consistent structure, familiar characters, and a single sense of style. On the other hand, it needs a variety of new events, some fresh situations and people, and the occasional surprise. But drama can also be used in the one-off situation to make a specific point, for example in describing how a shopper should compare two supermarket products in order to arrive at a best buy. It can be far more effective than a talk by an official, however expert he might be. Lively, colloquial, simple dialogue, using two or three voices with effects behind it to give it location – in a store or factory, on a bus or at the hospital – this can be an excellent vehicle for explaining legislation affecting citizen rights, a new medical technique, or for providing background on current affairs. Scriptwriters, however, will recognise in the listener an immediate rejection of any expository material which has about it the feel of propaganda. The most useful ingredients would appear to be: ordinary everyday humour, credible characters with whom the listener can identify, profound scepticism and demonstrable truth.

Programmes for children use drama to tell stories or explain a point in the educational sense. The use of this form may involve separate production effort, but it can be none the less effective by being limited to the dramatised reading of a book, poem or historical document.

Item order

Having established the programme structure, set the overall style and decided on the treatment of each individual item, the actual order of the items can detract from or enhance the final result.

In the case of a traditional circus or variety performance in a theatre, the best item – the top of the bill – is kept until last. It is safe to use this method of maintaining interest through the show since the audience is largely captive and it underwrites the belief that whatever the audience reaction, things can only get better. With radio, the audience is anything but captive and needs a strong item at the beginning to attract the listener to the start of the show, thereafter using a number of devices to hold his interest through to the end. These are often referred to as 'hooks' – remarks or items designed to capture and retain the listener's interest. It is crucial to engage the audience's attention early in the programme – an enticing, imaginatively written menu is one way of doing this – but then to continue to reach out, to appeal to the listener's sense of fun, curiosity, or need. Of course this applies to all programmes, but the longer strip programming is particularly susceptible to becoming bland and characterless. There should always be something coming up to look forward to.

The news magazine will probably start with its lead story and gradually work through to the less important. However, if this structure is rigidly applied the programme becomes less and less interesting – what has been called a 'tadpole' shape. News programmes are therefore likely to keep items of known interest until the end, e.g. sport, stock markets and weather, or at least end with a summary for listeners who missed the opening headlines. Throughout the programme as much use as possible will be made of phrases like 'more about that later'. Some broadcasters deliberately avoid putting items in descending order of importance in order to keep 'good stuff for the second half'; a dubious practice if the listener is to accept that the station's editorial judgement is in itself something worth having. A better approach is to follow a news bulletin with an expansion of the main stories in current affairs form.

The news magazine item order will be dictated very largely by the editor's judgement of the importance of the material, while in the tightly structured general magazine the format itself may leave little room for manoeuvre. In the more open sequence other considerations apply and it is worth noting once again the practice of the variety music bill. If there are two comedians they appear in separate halves of the show, something breathtakingly exciting is followed by something beautiful and charming, the uproariously funny is complemented by the serious or sad, the small by the visually spectacular. In other words, items are not allowed simply to stand on their own but through the contrast of their own juxtaposition and the skill of the compère they enhance each other so that the total effect is greater than the sum of the individual parts.

So it should be with the radio magazine. Two interviews involving men's voices are best separated. An urgently important item can be heightened by something of a lighter nature. A long item needs to be followed by a short one. Women's voices, contributions by children or old people should be consciously used to provide contrast and variety. Tense, heated or other deeply felt situations need special care, for to follow with something too light or relaxed can give rise to accusations of trivialisation or superficiality. This is where the skill of the presenter counts for it is his or her choice of words and tone of voice which must adequately cope with the change of emotional level.

Variations in item style combined with a range of item treatment create endless possibilities for the imaginative producer. A programme which is becoming dull can be given a 'lift' with a piece of music half-way through, some humour in the links, or an audience participation spot towards the end. For a magazine in danger of 'seizing up' because the items are too long, the effect of a brief snippet of information in another voice is almost magical. And all the time the presenter keeps us informed on what is happening in the programme, what we are listening to and where we are going to next, and later. The successful magazine will run for years on the right mixture of consistency of style and unpredictability of content. It could be that apart from its presenter the only consistent characteristic is its unpredictability.

Examples

The following examples of the magazine format are not given as ideals but as working illustrations of the production principle. Commercial advertising has been omitted in order to show the programme structure more clearly but the commercial station can use its breaks to advantage, providing an even greater variety of content.

Example 1: Fortnightly half-hour industrial magazine

Structure	Running order	Actual timing
Standard opening (1′ 15″)	Signature tune	0 15
	Introduction	10
	Menu of content	15
	Follow-up to previous programmes	35
News (5 mins)	News round-up	5 05
	Link	15
Item	Interview on lead news story	3 08
	Link - information	30
Item	Voice piece on new process	1 52
	Link	15
Item	Vox pop - workers' views of safety rules	1 15
	Link	20
Trades Council spot ($2\frac{1}{2}$ mins)	Union affairs - spokesman	2 20
	Link	20
Participation spot (3 mins)	Listeners' letters	2 45
	Link - introduction	20
Discussion	Three speakers join presenter for discussion of current issue - variable length item to allow programme to run to time	6 20
Financial news (3 mins)	Market trends	3 00
Standard closing (50″)	Coming events	30
	Expectations for next programme	10
	Signature tune	10
		29′50″

In Example 1, the programme structure allows 1'15" for the opening and 50" for the closing. Other fixed items are a total of 8'00" for news, 3'00" for letters and 2'30" for the Union spot. About 2'00" are taken for the links. This means that just over half the programme runs to a set format leaving about 13'00" for the two or three topical items at the front and the discussion towards the end.

So long as the subject is well chosen the discussion is useful for maintaining interest through the early part of the programme since it preserves at least the possibility of controversy, interest and surprise. With a 'live' broadcast it is used as the buffer to keep things on time since the presenter knows he must bring it to an end and get into the 'Market trends' four minutes before the end of the programme. The signature tune is 'prefaded to time' to make the timing exact. With a recorded programme the discussion is easily dropped in favour of a featurette which in this case might be a factory visit.

Example 2: Weekly 25-minute religious current affairs

Structure	Running order	Timing
Standard opening (15")	Introduction	0 05
	Menu of content	0 10
Item	Interview – main topical interest	3 20
	Link	30
Item	Interview – woman missionary	2 05
	Link	10
Music (3 mins)	Review of gospel record release	2 40
	Link – information	55
Item	Interview (or voice piece) – forthcoming convention	2 30
	Link	05
Featurette (7 mins)	Personality – faith and work	7 10
	Link	15
News (5 mins)	News round-up and What's On events	4 45
Standard closing (15")	Closing credits	15
		————
		24'55"
		————

In Example 2, with no signature tune, the presenter quickly gets to the main item. Although in this edition all the items are interviews, they are kept different in character and music is deliberately introduced at the mid-point. The opening and closing take half a minute and the other fixed spots are allocated another fifteen minutes. If the programme is too long, adjustment can be made either by dropping some stories from the news or by shortening the music review. If it under-runs, a repeat of some of the record release can make a useful reprise at the end.

Example 3: Outline for a daily 2-hour afternoon sequence

Fixed times	Running order	Approximate durations
2.00	Sig. tune/intro-programme information/sig. tune	1 min
	Music –	2
	Item – human interest interview	3
	Quiz competition (inc. yesterday's result)	3
	Music –	3
2.15	Listeners' letters	5
	Music –	3
	Voice piece – background to current affairs	2
	Music –	2
	Leisure spot - home improvements, gardening	3
	Humour on record	2
	Studio discussion - topical talking point	10
	Music –	3
	"Out and about" spot - visit to place of interest	5
	Music – (non vocal)	2
3.00	News summary, sport and weather	2
	Phone-in spot	15
	Music –	3
	Film/theatre/TV review - coming events voice piece	4
	Music –	3
3.30	Special guest - interview	10
	Music - (illustrative)	2
	Item - child care or medical interview	3
	Quiz result	2
	Music – (non vocal)	2
3.50	Serial story - dramatised reading	9
	Closing sequence; items for tomorrow, production credits, sig. tune	1

In Example 3 the speech/music ratio is regulated to about 3:1. A general pattern has been adopted that items become longer as the programme proceeds. Each half of the programme contains a 'live' item which can be 'backtimed' to ensure the timekeeping of the fixed spots. These are respectively the studio discussion and the special guest interview. Nevertheless the fixed spots are preceded by non-vocal music prefaded to time for absolute accuracy. Fifteen minutes is allowed for links. An alternative method of planning this running order is by the clock format illustrated on p. 189.

Production method

A regular magazine or sequence has to be organised on two distinct levels – the long term and the immediate. Long-term planning allows for anniversaries, 'one-off editions', the booking of guests reflecting special events, or the running of related spots to form a series across several programmes. On the immediate timescale, detailed arrangements have to be finalised for the next programmes.

In the case of the fortnightly or weekly specialist magazine of the type represented in Examples 1 and 2, it is a fairly straightforward task for the producer to make all the necessary arrangements in association with the presenter and a small group of contributors acting as reporters or interviewers. The newsroom or sources of specialist information will also be asked for a specific commitment. The important thing is that everyone knows his brief and his deadline. It is essential too that while the producer will make the final editorial decisions, all the contributors feel able to suggest ideas. Good ideas are invariably 'honed up', polished and made better by the process of discussion. Suggestions are progressed further and more quickly when there is more than one mind at work on them. Almost all ideas benefit from the process of counter-suggestion, development and resolution. Lengthy programmes of this nature are best produced by team-working with clear leadership. This should be the pattern encouraged by the producer at the weekly planning meeting, by the end of which everyone should know what they have to do by when, and with what resources.

The daily sequence illustrated by Example 3 is more complex and such a broadcast will require a larger production team. Typically, the main items such as the serial story and special guest in the second half will have been decided well in advance but the subject of the discussion in the first may well be left until a day or two beforehand in order to reflect a topical issue. A retrospective look at something recently covered, the further implications of yesterday's story, a 'whatever happened to ...?' spot – these are all part of the daily programme. Responsibility for collating the listeners' letters, organising the quiz, and producing the 'Out and about' item will be delegated to specific individuals, and the newsroom will be made responsible for the current affairs voice piece, the news, sport and weather package, and perhaps one of the other interviews. The presenter will write his own links and probably choose his own music to a general brief from the producer. The detail is regulated at a morning planning meeting, the final running order being decided against the format structure which serves as the guideline.

At less frequent intervals, say once a month, the opportunity should be taken to stand back to review the programme and take stock of the long-term options. In this way it may be possible to prevent the onset of the more prevalent disease of the regular programme – getting into a rut – while at the same time avoiding the disruptive restlessness which results from an obsession with change. When you have become so confident about the success of the programme that you fail to innovate, you have already started the process of staleness which leads to failure. A good approach is constantly to know what you are up against. Identify your competition and ask 'What must I do to make my programme more distinguishable from the rest?'

Responding to emergency

Sooner or later all the careful planning has to be discarded in order to cope with emergency conditions in the audience area. Floods, snowfall, hurricane, earthquake, power failure, bush fire or major accident will cause a sequence to cancel everything except the primary reason for its existence – public service. It will provide information to isolated villages, link helpers with those in need, and act as a focus of community activity. 'Put something yellow in your window if you need a neighbour to call.' 'Here's how to attract the attention of the rescue helicopter which will be in your area at 10 o'clock . . .' 'The following schools are closed . . .' 'The army is opening an emergency fuel depot at . . .' 'A spare generator is available from . . .' 'Can anyone help an elderly woman at . . .?' The practical work that radio does in these circumstances is immense, its morale value in keeping isolated, and sometimes frightened, people informed is incalculable. Broadcasters are glad to work for long hours when they know that their service is uniquely valued.

The flexibility offered by the sequence format enables a station to cover a story, draw attention to a predicament, or devote itself totally to audience needs as the situation demands. Radio services should have their contingency plans continuously updated – at least reviewed annually – in order to be able to respond quickly to any situation. Programmers cannot afford to wait for an emergency before making decisions but must always be asking the question – 'what if . . .?'

16

Outside broadcasts (remotes)

As has been noted elsewhere, there is a tendency for broadcasters to shut themselves away in studios being enormously busy making programmes which do not originate from a direct contact with the audience. The outside broadcast, or 'remote', represents more than a desire to include in the schedule coverage of outside events in which there is public interest. It is a positive duty for the broadcaster to escape from the confines of his building into the world which is both the source and the target for all his enterprise. The concert, church service, exhibition, civic ceremony, sporting event, public meeting, conference or demonstration; these demand the broadcaster's attention. But it is not only good for radio to reflect what is going on, it is necessary for the station's credibility to be involved in

Figure 16.1 The radio station must involve itself in its own community, otherwise it will appear irrelevant and out of touch

such things. Radio must not only go to where people are, it must come from the interests and activities of many people. If its sources are too few, it is in danger of appearing detached, sectional, elitist or out of touch. Thus the OB is essential to broadcasting's health.

Planning

The producer in charge, together with the appropriate engineering staff, must first decide how much coverage is required of a specific event. The programme requirement must be established and the technical means of achieving it costed. Is it to be 'live', or recorded on-site? What duration is expected? Once there is a definite plan, the resources can be allocated – people, facilities, money and time.

It is also at this first stage that discussions must take place with the event organiser to establish the right to broadcast. It may be necessary to negotiate any fees payable, or conditions or limitations which he may wish to impose.

Visiting the site

A reconnaissance is essential, but it may take considerable imagination to anticipate what the actual conditions will be like 'on the day'. There are a number of questions which must be answered:

1   Where and of what type are the mains electricity supply points? Is the supply correctly earthed? Do I need my own battery or generator power?
2   Where is the best vantage point to see the most action? Will there have to be more than one?
3   Will the sound mixing be done in the building, or in a radio OB vehicle outside?
4   What on-site communications are required?
5   How many microphones and what type will be needed?
6   How long are the cable runs?
7   Will a public address system be in use? If so, where are the speakers?
8   What else will be present on the day? e.g. flags which obscure the view, vehicles or generators which might cause electrical interference, background music, other broadcasters.
9   What are the potential hazards and safety requirements?

Communications to base

If the programme is 'live', how is the signal to be sent to the controlling

Site plan

Location THE CHURCH HALL, HIGH STREET, EXLEY

Contact name JOHN SMITH   **Post** Caretaker   **Phone** 135531

Transmission   **Date** 1 JULY   **Time** 1900–2000

Rehearsal   **Date** 29 JUNE   **Time** 2000–2230

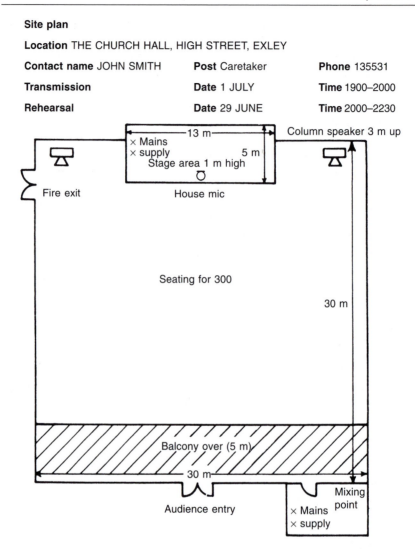

Figure 16.2 and label text:
- 13 m
- × Mains
- × supply
- Stage area 1 m high
- 5 m
- Column speaker 3 m up
- Fire exit
- House mic
- Seating for 300
- 30 m
- Balcony over (5 m)
- 30 m
- Audience entry
- × Mains
- × supply
- Mixing point

Figure 16.2 A sketch plan drawn up during the site visit is an invaluable aid to further planning

studio? What radio links are required? Is the site within available radio car range? Are land lines available? They may be expensive but will additional programme or control circuits need to be ordered from the Telecommunications Department? If so will the quality be good enough for music – or will the programme circuit have to be 'equalised'? These questions need to be discussed at an early stage, for, amongst other things, the answers will have a direct bearing on the cost of the programme.

Sooner or later an OB will be required from a hall, or from the middle of a field, where no lines exist. Such conditions are best realised well in advance so that either a radio-link can be obtained or the telephone authority asked to make the appropriate connections, and if necessary actually build a suitable route. A decision has to be made as to whether only a one-way programme circuit is required – the broadcaster at the site must then be able to obtain his cue to go ahead by listening off-air – or whether a second two-way telephone or control line is also needed. Obviously this additional facility is to be preferred and for an OB of any length and where a number of broadcasts are made from the same site, it becomes essential. The same applies if radio links are used – is there to be a bi-directional control channel in addition to the programme circuit from OB to base? The specialist producer will certainly know about the mobile phone and ISDN options described earlier on p. 105. If all else fails, however, it is a wise precaution to know the whereabouts of the nearest accessible telephone, whether it is in a private house or office, or a public call box. Many a programme has been saved by having the right coins available!

People

By this stage it should be clear as to how many people will be involved at the OB site. Anything more than a simple radio car job may require a number of skills – producer, engineer, commentator, technical operator, secretary, driver, caterer, etc. A large event with the public present, such as an exhibition, may require the services of security staff, or a publicity specialist. The list grows with the complexity of the programme, as does the cost.

The exact number of people is an important piece of anticipation – getting it right depends on being able to visualise whether, for example, there is a script writing and typing requirement on-site. It will also depend on whether the working day is to be so long as to warrant the employment of duplicate staff working in shifts.

Hazard assessment

Away from the familiarity of the home studio, OBs are full of uncertainties. During the reconnaissance visit there must be a careful assessment of the potential hazards, and for the safety of staff, performers, and members of the public, plans made to minimise them. This applies not only to the broadcast but also to the rigging and derigging before and afterwards.

What measures are needed for crowd control – especially children? Are there any dangers from water, fire, heights, vehicles, traffic, animals, aircraft, etc.? Are hard hats, fluorescent jackets, or any protective clothing needed? What contingency plans exist for dealing with an emergency? Should the police or other authorities be informed? No activity of this nature is entirely without risk, but responsible steps must be taken to anticipate and avoid possible accidents.

Equipment

This is best organised on a category basis by the individuals most closely involved:

1 *Engineering:* microphones including radio mics, cables, leads and connectors, audio mixers, computer kit, TV video link, cassette or CD decks, recorders, amplifiers, loudspeakers, headphones, recording and editing materials, digital Fx unit, radio, spare batteries, power cables and distribution boards, isolating transformers, circuit breakers, spare fuses, heavy-duty sticky tape, tool kit, fire extinguisher.
2 *Programme:* CDs, tapes, discs, scripts, stopwatches.
3 *Administrative:* tables, chairs, paper, laptop or notebook, printer, pens and pencils, marker pens, torches, clipboards, money, string, rope, publicity and sign-writing materials, sticky tape, mobile phone.
4 *Personal:* food and drink, special clothing, first aid kit, sleeping bags, etc.
5 *Transport:* vehicles, maps, shovel.

There are always things that get forgotten, but if they are really important, this only happens once.

Accommodation

In further discussion with the event organiser there has to be agreement on the exact location of the broadcasting personnel and equipment. There may be special regulations governing car parking or access to the site, in which case the appropriate passes and security badges need to be obtained.

In order to rig equipment it will be necessary to gain access well before the event – are any keys required? Who will be there? and what security exists to safeguard equipment once it is rigged? At this stage the producer must also satisfy himself as to the whereabouts of lavatories, fire exits, catering facilities, lifts and any special features of the site, e.g. steps, small doorways, awkward passages, non-opening windows, or unusual acoustics.

The audio mixer frequently remains outside in the OB vehicle or is in a room from which the action cannot be seen. Under these conditions it is extremely useful to have a TV camera feeding a monitor adjacent to the mixing desk so that the event can be followed – and anticipated, by the producer and engineer. The security and siting of the camera should be agreed with the event organiser in advance.

Programme research

Further discussion with the event organiser will establish the detailed timetable and list of participants. With an open-air event such as a parade or sports meeting, it is important to discover any alternative arrangements in case of rain. As much information as possible about who is taking part, the history of the event, how many people have attended on previous occasions, and so on, is useful preparatory material for the broadcast itself. Additional research may be necessary at this stage for the commentators – see under 'Commentary' (Chapter 17).

The producer is then in a position to draw up a running order and to tell everyone his or her precise role both on and off the air. The running order should give as much relevant information as possible including who is doing what, when, and details of cues and timings, where these are known.

Liaison with the base studio

Particularly in the case of a 'live' OB, staff at the base studio need to be kept informed. They should have copies of the running order and be involved in a discussion of any special fill-up material or other instructions in case of a technical failure. Arrangements should be made for a 'live' OB to be recorded at the base studio so that a possible feature can be made for a highlight or follow-up programme.

Publicity

The producer should ensure he has done everything he can to provide the appropriate advance publicity. This may be in the form of programme billing, printed posters, a press release, or simply on-air trails and announcements. It is a matter of common experience that broadcasters go to immense pains to cover an important public event, but overlook the necessity of telling people about it in advance. It is true that some promoters claim that broadcast coverage keeps people away from attending the event itself, but on the other hand, it is frequently the case that advance publicity will stimulate public interest and swell the crowds, many of whom will follow the action via radio.

Safety

In a situation where crowds of people are present and their attention is inevitably drawn to the spectacle they have come to see, the broadcaster has a special responsibility to ensure that his own operation does not present any hazards. The broadcaster is of course affected by, and must observe, any local bye-laws or other regulations which apply to the OB

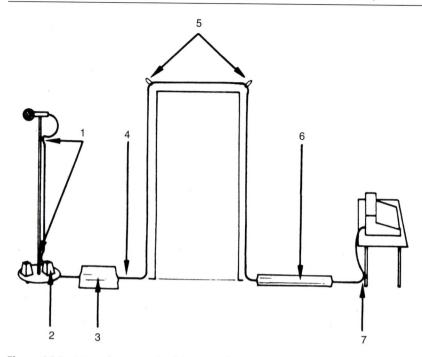

Figure 16.3 Microphones and cables must be safely rigged. 1. Cable taped to mic stand. 2. Stand weighted. 3. Mat covers cable at walkway. 4. Cable laid without loops or knots. 5. Cable slung over doorways. 6. Cable taped to floor. 7. Cable made fast at mixer end

site. His equipment must not obstruct gangways or obscure fire exit notices, or the fire equipment itself.

Cables across pavements or passage-ways must either be covered by a ramp, or should be lifted clear of any possibility of causing an obstruction.

Microphones suspended over an audience must be securely fixed, not just with sticky tape which can become loosened with a temperature change, but secured in such a way as to prevent any possibility of their being untied by inquisitive or malicious fingers. Safety chains should doubly secure any equipment rigged overhead.

Members of the public are generally curious of the broadcasting operation and all equipment must be completely stable, for instance microphone stands or loudspeakers should not be able to fall over. Nor of course must straying hands be able to touch mains electricity connections. A little fencing off may be necessary.

This will almost certainly be the case where a ground-level public address system is in use. For a DJ show where the broadcaster is providing an on-site loudspeaker output, the sound intensity close to the speakers is often sufficient to cause temporary, and in some cases permanent, damage to ears. To prevent this some form of barrier three metres or so from the speaker is generally needed. A better alternative is to raise the speakers, fixing them securely well above head height.

Conflicts of approach

Engineers are essential to the broadcasting process and programme people ought to know that they wouldn't get far without them – especially at an outside broadcast. While producers might well be criticised for taking too little interest in technical matters, it also has to be said that some engineers have a tendency to over-complicate things. They can use far too many mics, or have an undue liking for technical gimmicks, and they do not always explain to the producer, as fully as they might, the problems and possibilities of the situation. Furthermore the mixing desk is often operated by a technician or specialist OB engineer and so the final mix – the programme sound – is not under the producer's direct control. So while an engineer at an OB may be aiming at studio speech quality and acoustically perfect sound, the producer may want much more the *event* – a sense of the occasion. Because of such differences of approach it is not surprising that arguments arise between producers and engineers. The problem is generally compounded by the stressful pressure of time.

A common difficulty with the broadcasting of live events is the frequent lack of any proper rehearsal. It is often not possible to listen to the mix, discuss it, make adjustments, and do it again. It is therefore essential that the producer discusses his or her objectives with the balance engineer well in advance and maintains close contact to resolve problems as they arise. Confronted by an engineer of many years' experience the young producer is likely to feel daunted by the older person and unable to question anything said or done. By all means use that experience, depend on that competence and learn from it – but also develop the personal skills to challenge precedent and make changes if you feel it right. After all, in most broadcasting organisations it is the producer who is in charge – and responsible for the end result.

Tidiness

The broadcaster is working in a public place and both his appearance and his general behaviour will contribute to the station's image. This is recognised by the more senior staff but may not always be appreciated by freelance contributors. A small but important aspect of public relations is the matter of leaving the OB site in a sensibly tidy state. It is clearly undesirable to leave any equipment behind but this should also apply to the accumulated rubbish of a working visit – scripts, notes, food tins, plastic bags, empty boxes, etc. To be practical, a good OB site will be required again and it is not in the broadcaster's interests to be remembered for the wrong reasons.

Gratuities

It is common sense to recognise that the broadcaster's presence at an OB site is likely to cause extra effort for those who normally work there. It will not be necessary to consider this point in every case, and it may even be felt that where there is some special distinction attached to broadcasting it is possible to live off that good name. This temptation should be resisted, for every time that status is used in this way, it is likely to be diminished. A facility fee should be paid where local assistance is provided beyond the normal level, and to any outsiders who supplied some special service, for example the use of a telephone, electricity or water, or the parking of vehicles. The amounts should obviously be related to the service provided – too much and one is open to charges of profligate wastage, too little and one quickly does more harm than good.

17

Commentary

Radio has a marvellous facility for creating pictures in the listener's mind. It is more flexible than television in that it is possible to isolate a tiny detail without waiting for the camera to 'zoom in' and it can create a breadth of vision much larger than the dimensions of a glass screen. The listener does more than simply eavesdrop on an event; radio, more easily than television or film, can convey the impression of actual participation. The aim of the radio commentator is therefore to recreate in the listener's mind not simply a picture but a total impression of the occasion. This is done in three distinct ways:

1 The words used will be visually descriptive of the scene.
2 The speed and style of their delivery will underline the emotional mood of the event.
3 Additional 'effects' microphones will reinforce the action, or the public reaction to it.

Attitude to the listener

In describing a scene the commentator should have in mind 'a blind friend who couldn't be there'. It is important to remember the obvious fact that the listener cannot see. Without this it is easy to slip into the situation of simply chatting about the event to 'someone beside you'. The listener should be regarded as a friend because this implies a real concern to communicate accurately and fully. The commentator must use more than his eyes and convey information through all the senses, so as to heighten the feeling of participation by the listener. Thus, for example, temperature, the proximity of people and things, or the sense of smell are important factors in the overall impression. Smell is particularly evocative – the scent of newly mown grass, the aroma inside a fruit market or the timeless mustiness of an old building. Combine this with the appropriate style of delivery, and the sounds of the place itself, and you are on the way to creating a powerful set of pictures.

Preparation

Some of the essential stages are described under 'Outside broadcasts' but the value of a pre-transmission site visit cannot be over-emphasised. Not only must the commentator satisfy himself as to his field of vision and whether the sun is in his eyes, but he must use the time to obtain essential facts about the event itself. For example, in preparing for a ceremonial occasion he will need to discover:

1 The official programme of events with details of timing, etc.
2 The names of the flowers used for decoration, or the trees in the area.
3 The history of the buildings and streets, and their architectural detail.
4 The background of the people taking part – unseen as well as seen, for example organisers, caretakers.
5 The titles of music to be played, and any special association it may have with the people and the place.

It adds immeasurably to the description of a scene to be able to mention the type of stonework used in a building, or that 'around the platform are fuchsias and hydrangeas'. The point of such detail is to use it as contrast with the really significant elements of the event so letting them gain in importance. Contrast makes for variety and for more interesting listening and mention of matters both great and small is essential, particularly for an extended piece. An eye for detail can also be the saving of a broadcast when there are unexpected moments to fill. There is no substitute for a commentator doing his homework.

In addition to personal observation and enquiry, useful sources of information will be the reference section of libraries or museums, newspaper cuttings, back copies of the event programme, previous participants, specialist magazines, and commercial or government press offices.

Having obtained all the factual information in advance, the commentator must assemble it in a form which he can use in the conditions prevailing at the OB. If he is perched precariously on top of a radio car in the rain, clutching a guardrail to steady himself and holding a microphone, stopwatch, and an umbrella, the last thing he wants is a bundle of easily windblown papers! Notes should be laid out on as few sheets as possible and held firmly on a clipboard. Cards may be useful since they are silent to handle. The important thing is their order and logic. The information will often be chronological in nature listing the background of the people taking part. This is particularly so where the participants appear in a predetermined sequence – a procession or parade, variety show, race meeting, athletics event, church service, or civic ceremony. Further information on the event or the environment can be on separate pages so long as they can be referred to easily. If the event is non-sequential, for instance a football match or public meeting, the personal information may be more useful in alphabetical form, or better still memorised.

Working with the base studio

The commentator will need to know the precise handover details. This applies both from the studio to himself and from his own end cue for the return to the studio. These details are best written down for they easily slip the memory. He should also be fully aware of the procedures to be followed in the event of any kind of circuit failure – the back-up music to be played, and who makes the decision to restore the programme. It may be necessary to devise some system of hand signals or other means of communication with technical staff, and he will want to know whether he will be able to hear 'talkback' in his headphones, etc. These matters are the 'safety nets' which enable the commentator to fulfil his role with a proper degree of confidence.

As with all outside broadcasts, the base studio should ensure that the commentary output is recorded. Not only will the commentator be professionally curious as to how it came over, but the material may be required for archive purposes. Even more important is that an event worthy of a 'live' OB will almost certainly merit a broadcast of edited highlights later in the day. Thus two recordings may be needed, a complete original and one which can be cut for subsequent rebroadcast.

Sport

First and foremost the sports commentator must know his or her sport and have detailed knowledge of the particular event. He should know the sequence which led up to it, its significance in any overall contest, the participants and something of their history. The possession of this background information is elementary, but what is not so obvious is how to use it. The tendency is to give it all out at the beginning in the form of an encyclopedic but fairly indigestible introduction. Certainly the basic facts must be provided at the outset, but a much better way of using background detail is as the game itself proceeds, at an appropriate moment or during a pause in the action. This way the commentator sounds as though he is part of what is going on instead of being a rather superior observer.

Traditionally, for technical reasons, the commentator has often had to operate from inside a soundproof commentary position, isolated from his immediate surroundings. He can easily lose something of the atmosphere by creating his own environment and there is a strong argument in favour of the ringside seat approach, provided that he uses a noise-cancelling microphone and his communication facilities, such as headphone talkback, are secure.

Sports stadia seem to undergo more frequent changes to their layout than other buildings and unless a particular site is in almost weekly use for radio work, a special reconnaissance visit is strongly advised. It is easy to forego the site reconnaissance assuming that the place will be the

Figure 17.1 The lip mic. The microphone has excellent noise-cancellation properties which makes it ideal for commentary situations. The mouthguard is held against the broadcaster's lip while the microphone is in use. There is a bass cut in the handle to compensate for the bass lift which results from working close to a ribbon microphone

same as it was six months ago. However, unless there are strong reasons to the contrary, a visit and technical test should always be made.

Where the action is spread out over a large area as with motor racing, a full-scale athletics competition or rowing event, more than one commentator is likely to be in action. Cueing, handovers, timing, liaison with official results – all these must be precisely arranged. The more complex an occasion, the more necessary is observance of the three golden rules for all broadcasts of this type:

1 Meticulous production planning so that everyone knows what is *likely* to be asked of him.
2 First-class communications for control.
3 Only one person in charge.

Communicating mood

The key question is 'what is the overall impression here?' Is it one of joyful festivity or is there a more intense excitement? Is there an urgency to the occasion or is it relaxed? At the other end of the emotional scale there may be a sense of awe, a tragedy or a sadness which needs to be reflected in a sombre dignity. Whatever is happening, the commentator's sensitivity to its mood, and to that of the spectators, will control his style, use of words and speed of delivery. More than anything else this will carry the impressions of the event in the opening moments of the broadcast. The mood of the crowd should be closely observed – anticipatory, excited, happy, generous, relaxed, impressed, restive, sullen, tense, angry, solemn, sad. Such feelings should be conveyed in the voice of the commentator and their accurate assessment will help him to know when to stop and let the sounds of the event speak for themselves.

Coordinating the images

It is all too easy to fall short of an overall picture but to end up instead with some accurately described but separate pieces of jigsaw. The great art, and challenge, of commentary is to fit them together, presenting them in a logically coordinated way which allows your 'blind friend' to place the information accurately in his mind's eye. The commentator must include not only the information relating to the scene, but also something about how this information should be integrated to build the appropriate framework of scale. Having provided the context, other items can then be related to it. Early on, it should be mentioned where the commentator's own position is relative to the scene; also giving details of distance, size, foreground, left and right, etc. Movement within a scene needs a smooth, logical transition if the listener is not to become hopelessly disorientated.

Content and style

The commentator begins with a 'scene-set'. He says first of all where he is and why – this is best not given in advance by the continuity handover and duplication of this information must be avoided. The listener should be helped to identify with the location, particularly if it is likely to be familiar to him. The description continues from the general to the particular noting, as appropriate, the weather, the overall impression of lighting, the mood of the crowd, the colour content of the scene and what is about to happen. Perhaps two minutes or more should be allowed for this 'scene-setting', depending on the complexity of the event, during which time nothing much may be happening. By the time the action begins the listener should have a clear visual and emotional picture of the setting, its sense of scale and overall 'feel'. Even so, the commentator must continually refer to the generalities of the scene as well as to the detail of the action. The two should be woven together.

Time taken for scene-setting does not of course apply in the case of news commentary where one is concerned first and foremost with what is happening. Arriving at the scene of a fire or demonstration, one deals first with the event and widens later to include the general environment. Even so, it is important to provide the detail with its context.

Many commentaries are greatly improved by the use of colour. Colour

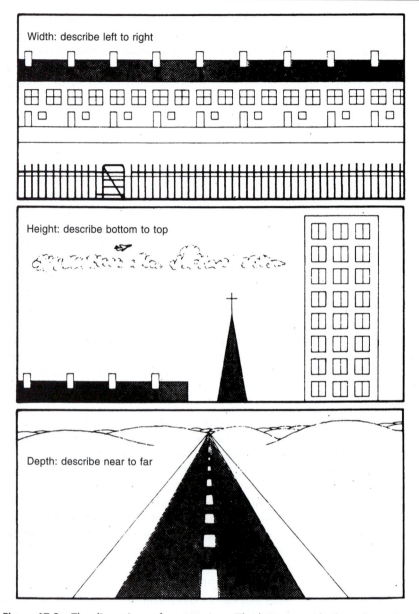

Figure 17.2 The dimensions of commentary. The listener needs three-dimensional information in which to place the action. Such orientation should not be confined to the scene-set but should be maintained throughout the commentary

whether gaudy or sombre is easily recreated in the mind's eye and mention of purple robes, brilliant green plumage, dark grey leaden skies, the blue and gold of ceremonial, the flashes of red or the sparkling white surf – such specific references conjure up the reality much better than the short cut of describing a scene simply as multi-coloured.

In describing the action itself, the commentator must proceed at the same pace as the event, combining prepared fact with spontaneous vision. In the case of a planned sequence as a particular person appears, or slightly in anticipation, reference is made to the appropriate background information, title, relevant history, and so on. This is more easily said than done and requires a lot of practice – perhaps using a recorder to help perfect the technique.

Sports action

The description of sport, even more than that of the ceremonial occasion, needs a firm frame of reference. Most listeners will be very familiar with the layout of the event and can orient themselves to the action so long as it is presented to them the right way round. They need to know which team is playing from left to right; in cricket which end of the pitch the bowling is coming from; in tennis, who is serving from which court; in horse racing or athletics, the commentator's position relative to the finishing line. It is not sufficient to give this information at the beginning only, it has to be used throughout a commentary, consciously associated with the description of the action.

As with the ceremonial commentator, the sports commentator is keeping up with the action but also noticing what is going on elsewhere, for instance the injured player, or a likely change in the weather. Furthermore, the experienced sports commentator can increase interest for the listener by highlighting an aspect of the event which is not at the front of the action. For example, the real significance of a motor race may be the duel going on between fourth and fifth place; the winner of a 10,000 metre race may already be decided but there can still be excitement in whether the athlete in second place will set a new European record or a personal best.

With slower games, such as cricket, the art is to use pauses in the action interestingly, not as gaps to fill but opportunities to add to the general picture or give further information. This is where another commentator or researcher can be useful in supplying appropriate information from the record books or with an analysis of the performance to date. Long stretches of commentary in any case require a change of voice, as much for the listener as for the commentators, and changeovers every 20–30 minutes are about the norm.

If for some reason the commentator cannot quite see a particular incident or is unsure what is happening, he should avoid committing himself – 'I think that . . .' A better, more positive way is 'It looks as though . . .' Similarly it is unwise for a commentator to speculate on what a referee is saying to a player in a disciplinary situation. Only what can be seen or known should be described. It is easy to make a serious mistake affecting an individual's reputation through the incorrect interpretation of what may appear obvious. And it must be regarded as quite exceptional to voice a positive disagreement with a referee's decision. After all he is closer to the action and may have seen something which the commentator, with his general view, missed. The reverse can also be true but in the heat of the moment it is sensible policy to give referees and umpires the benefit of any doubt.

Scores and results should be given frequently for the benefit of listeners who have just switched on but in a variety of styles in order to avoid irritating those who have listened throughout. Commentators should remember that the absence of goals or points can be just as important as a positive scoreline.

Actuality and silence

It may be that during the event there are sounds to which the commentary should refer. The difficulty here is that the noisier the environment, the closer on-microphone will be the commentator so that the background will be relatively reduced. It is essential to check that these other sounds can be heard through separate microphones otherwise references to 'the roar of the helicopters overhead', 'the colossal explosions going on around me' or 'the shouts of the crowd' will be quite lost on the listener. It is important in these circumstances for the commentator to stop talking and to let the event speak for itself.

There may be times when it is virtually obligatory for the commentator to be silent – during the playing of a national anthem, the blessing at the end of a church service, or important words spoken during a ceremonial. Acute embarrassment on the part of the over-talkative commentator and considerable annoyance for the listener will result from his being caught unawares in this way. A broadcaster unfamiliar with such things as military parades or church services must be certain to avoid such pitfalls by a thorough briefing beforehand.

The ending

Running to time is helped by having a stopwatch synchronised with the studio clock. This will provide for an accurately timed handback, but if open-ended, the cue back to the studio is simply given at the conclusion of the event.

It is all too easy after the excitement of what has been happening to create a sense of anti-climax. Even though the event is over and the crowds are filtering away, the commentary should maintain the spirit of the event itself perhaps with a brief summary, or with a mention of the next similar occasion. Another technique is radio's equivalent of the television wide-angle shot. The commentator 'pulls back' from the detail of the scene, concluding as he began with a general impression of the whole picture before ending with a positive and previously agreed form of words which indicates a return to the studio.

Many broadcasters prefer openings and closings to be scripted. Certainly if you have hit upon the neat, well-turned phrase, its inclusion in any final paragraph will contribute appropriately to the commentator's endeavour to sum up both the spirit and the action of the hour.

An example

One of the most notable commentators was the late Richard Dimbleby of the BBC. Of many, perhaps his most memorable piece of work was his description of the lying-in-state of King George VI at Westminster Hall in February 1952. The printed page can hardly do it justice, it is radio and should be heard to be fully appreciated. Nevertheless it is possible to see here the application of the commentator's 'rules'. A style of language, and delivery, that is appropriate to the occasion. A 'scene-set' which quickly establishes the listener both in terms of the place and the mood. 'Signposts' which indicate the part of the picture being described. Smooth transitions of movement which take you from one part of that picture to another. Researched information, short sentences or phrases, direct speech, colour and attention to detail, all used with masterly skill to place the listener at the scene.

'It is dark in New Palace Yard at Westminster tonight. As I look down from this old, leaded window I can see the ancient courtyard dappled with little pools of light where the lamps of London try to pierce the biting, wintry gloom and fail. And moving through the darkness of the night is an even darker stream of human beings, coming, almost noiselessly, from under a long, white canopy that crosses the pavement and ends at the great doors of Westminster Hall. They speak very little, these people, but their footsteps sound faintly as they cross the yard and go out through the gates, back into the night from which they came.

They are passing, in their thousands, through the hall of history while history is being made. No one knows from where they come or where they go, but they are the people, and to watch them pass is to see the nation pass.

It is very simple, this lying-in-state of a dead king, and of incomparable beauty. High above all light and shadow and rich in carving is the massive roof of chestnut that Richard II put over the great hall. From that roof the light slants down in clear, straight beams, unclouded by any dust, and gathers in a pool at one place. There lies the coffin of the King.

The oak of Sandringham, hidden beneath the rich golden folds of the Standard; the slow flicker of the candles touches gently the gems

of the Imperial Crown, even that ruby that King Henry wore at Agincourt.
It touches the deep purple of the velvet cushion and the cool, white
flowers of the only wreath that lies upon the flag. How moving can
such simplicity be. How real the tears of those who pass and see it, and
come out again, as they do at this moment in unbroken stream, to the
cold, dark night and a little privacy for their thoughts.'

(Richard Dimbleby)

Coping with disaster

Sooner or later something will go wrong. The VIP plane crashes, the
football stadium catches fire, terrorists suddenly appear, a peaceful
demonstration unexpectedly becomes violent, or spectator stands collapse.
The specialist war correspondent or experienced news reporter sent to
cover a disaster knows how far to go in describing death and destruction.
Sensitivity to the reactions of the listener in describing mutilated bodies
or the bloody effect of shellfire is a matter which is developed through
experience and a constant reappraisal of news values. But the non-news
commentator must also learn to cope with tragedy. From the crashing of
the Hindenburg airship in 1937, to the explosion of the space shuttle,
commentators are required to react to the totally unforeseen, responding
with an instant transition perhaps from national ceremony to fearful disaster.
Certain kinds of events such as motor racing and airshows have an inherent
capacity for accident, but when terrorists invade a peaceful Olympic
Games village commentators normally used to describing the excitement
of the track are called on to cope with tensions of quite a different magnitude.

Here is an example of BBC Sports commentator Peter Jones covering
a football match at Hillsborough Stadium, Sheffield. The game had only
just begun when more people crowding into the ground suddenly caused
such a crush in the stands that supporters were climbing the fences and
invading the pitch. The match was stopped and a few moments later:

'At the moment there are unconfirmed reports, and I stress unconfirmed
reports of five dead and many seriously injured here at Hillsborough.
Just to remind you what happened – after five minutes, at the end of the
ground to our left where the Liverpool supporters were packed very
tightly – and the report is that one of the gates in the iron fence burst
open – supporters poured on the pitch. Police intervened and quite
correctly the referee took the police advice and took both teams off.
Since then we've had scenes of improvised stretchers with the advertising
hoardings being torn up, spectators have helped, we've got medical
teams, oxygen cylinders, a team of fire brigade officers as well to
break down the fence at one end to make it easier for the ambulances
to get through and we've got bodies lying everywhere on the pitch.'

(Courtesy BBC Sport)

Remembering that his commentary was being heard by the families and friends of people at the match, it was important here to describe the early casualty reports as 'unconfirmed' and also to avoid any attempt at identifying the cause of the situation, or worse, to apportion blame. Commentators do well to report only what they can personally see. So what should the non-specialist do? Here are some guidelines:

- Keep going if you can. Don't be so easily deterred by something unusual that you hand back to the studio. Even if your commentary is not broadcast 'live' it could be crucial for later news coverage.
- There's no need to be ashamed of your own emotions. You are a human being too and if you are horrified or frightened by what is happening your own reaction will be part of conveying that to your listener. It's one thing to be professional, objective and dispassionate at a planned event, it is quite another to remain so during a sudden emergency.
- Don't put your own life, or the lives of others, in unnecessary danger. You may from the best of motives believe that 'the show must go on', but few organisations will thank you for the kind of heroics which result in your death. If you are in a building which is on fire, say so and leave. If the bullets are flying or riot gas is being used in a demonstration, take cover. You can then say what's happening and work out the best vantage point from which to continue.
- Don't dwell on individual anguish or grief. Keep a reasonably 'wide angle' and put what is happening in context. Remember the likelihood that people listening will have relatives or friends at the event.
- Let the sounds speak for themselves. Don't feel you have to keep talking, there is much value in letting your listener hear the actuality – gunfire, explosions, crowd noise, shouts and screams.
- Don't jump too swiftly to conclusions as to causes, and responsibility. Leave that to a later perspective. Stick with observable events, relay the facts as you see them.
- Above all arrive at a station policy for this sort of coverage well before any such event takes place. Get the subject on the agenda in order to agree emergency procedures.

18

Music recording

There are three questions which a producer must ask himself before becoming involved in the production of any music.

First, is the material offered relevant to programme needs? The technically minded producer, audiophile or engineer can easily create reasons for recording a particular music occasion other than for its value as good programming. It may represent an attractive technical challenge, or be the sort of concert which at the time seems a good idea to have 'in the can'. Alternatively the desire to record a particular group of performers may outweigh considerations of the suitability of what they are playing. Sometimes musicians are visually persuasive but aurally colourless, or as with many club performers dependent on an atmosphere difficult to reproduce on radio. To embark on music with little prospect of broadcasting it is a speculative business and the radio producer should not normally commit resources unless he knows, perhaps only in outline, how he intends to use the material.

The second question asks whether the standard of performance is good enough for broadcasting. There cannot be a single set of objective criteria for standards since much depends on the programme's purpose and context. A national broadcaster will undoubtedly demand the highest possible standards; regionally and locally there is an obligation to broadcast the music making of the area which will almost certainly be of less than international excellence. The city orchestra, college swing band, amateur pop group, all have a place in the schedules but for broadcasting to a general audience, as opposed to a school concert which is directed only to the parents of performers, there is a lower limit below which the standard must not fall. In identifying this minimum level, the producer has to decide whether he is primarily broadcasting the music, the musician, or the occasion. Certainly he should not go ahead without a clear indication of the likely outcome – new groups should be auditioned first, preferably 'live' rather than from a submitted tape. If the technical limits of performance are apparent, musicians should be persuaded to play items that lie within their abilities. Simple music well played is infinitely preferable to the firework display that does not come off.

And finally, is the recording or broadcast within the technical capability of the station? Even at national level there are limits to what can be expended on a single programme. Numbers and types of microphones, the best specification for stereo lines, special circuits, engineering facilities at a remote OB site, the availability of more than one echo source and so on – these are considerations which affect what the listener will hear and will therefore contribute to his appreciation of the performer. The broadcaster has an obvious obligation to artists to present them in the most appropriate manner without the intervention of technical limits. This presupposes that the producer knows exactly what is involved in any given situation, that, for example, he understands the implications of a 'live' concert where the members of a pop group sing as they play, where public address relay is present, or hypercardioid, variable pattern or tie-clip microphones are required to do justice to a particular sound balance. Small programme-making units should avoid taking on more than they can adequately handle, and instead should stay within the limits of their equipment and expertise. Much better for a local station to refuse the occasional music OB as beyond its technical scope than it should broadcast a programme which it knows could have been improved on given better equipment. Alternatively if a station is not equipped or staffed to undertake outside music recording, it could consider contracting the work out to a specialist facilities company, with the station retaining editorial control.

Choir – 34 performers = 1 mic

Pop Group – 4 performers = 12 mics

Figure 18.1 The technical complexity of broadcasting live music is not related to how many performers there are. The producer must decide whether adequate programme standards can be achieved within the available resources

The remainder of this chapter is directed to help producers in their understanding of the technical factors of music recording.

The philosophy of music balance is divided into two main groups – first, the reproduction of a sound which is already in existence, and second, the creation of a synthetic overall balance which exists only in the composer's or arranger's head and subsequently in the listener's loudspeaker.

Reproduction of internal balance

Where the music produced results from a carefully controlled and self-regulating relationship between the performers, it would be wrong for the broadcaster to alter what the musicians are trying to achieve. For example, the members of a string quartet are sensitive to each other and adjust their individual volume as the music proceeds. They produce a varying blend of sound which is as much part of the performance as the notes they play or the tempo they adopt. The finished product of the sound already exists and the art of the broadcaster is to find the place where his microphone(s) can most faithfully reproduce it. Other examples of internally balanced music are symphony orchestras, concert recitalists, choirs and brass bands. The dynamic relationship between the instruments and sections of a good orchestra is under the control of its conductor; the broadcaster's task is to reproduce this interpretation of the music. He should not create a new sound by boosting the woodwind or unduly accentuating the trumpet solo. Since the conductor controls the internal balance by what he hears, it is in this area that one searches initially for the 'right' sound. Using a 'one-mic balance' or stereo pair, the rule of thumb is to place it with respect to the conductor's head – 'three metres (10 feet) up and three metres (10 feet) back'.

Similarly the conductor of a good choir will on his own assessment regulate the balance between the soprano, contralto, tenor and bass parts. If the choir is short of tenors and this section needs reinforcing, this adjustment can of course be made in the microphone balance. But this is at once to create a sound not made simply by the musicians and raises interesting questions about the lengths to which broadcasters may go in repairing the deficiencies of performers.

It is possible to 'improve' a musician's tone quality, to clarify the diction of a choir or to correct an unevenly balanced group. The ultimate in such cosmetic treatment is to use the techniques of the recording studio to so enhance a performer's work that it becomes impossible for him ever to appear 'live' in front of an audience, a not-unknown situation. But while it is entirely reasonable to make every possible adjustment in order to obtain the best sound from a school orchestra, it would be unthinkable to tamper with the performance of an artist who had a personal reputation. It is in the middle ground, including the best amateur musicians, where the producer is required to exercise his judgement – to do justice to the artist but without making it difficult for the performer in other circumstances.

The foregoing assumes that the music is performed in a hall that has favourable acoustics. Where this is not the case, steps have to be taken to correct the particular fault. For example, in too 'dry' an acoustic, some artificial reverberation will need to be added, preferably on-site at the time of recording, although it is possible for it to be added later. Alternatively a 'space' microphone can be used to increase the pick-up of reflected sound. If the hall is too reverberant or the ambient noise level, from air-conditioning plant or outside traffic, is unacceptably high, the microphone should be moved closer to the sound source. If this tends to favour one

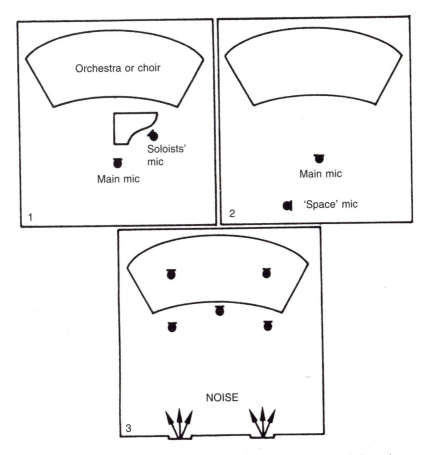

Figure 18.2 Basic considerations for mic placing with an internally balanced group. 1. Mic balance suitable under low noise conditions in a good acoustic. A soloist's mic or occasional 'filler' mic may be added. 2. In too 'dry' an acoustic a space mic is added to increase the pick-up of reflected sound. 3. Under conditions of unacceptable ambient noise the microphones are moved closer to the source and increased in number to preserve the overall coverage. It may be necessary to add artificial reverberation

part at the expense of another, further microphones are added to restore full coverage. The best sound quality from most instruments and ensembles is to be found not simply in front of them, but above them. A good music studio will have a high ceiling and the microphone for an internally balanced group will be kept well above head height. As mentioned previously, the standard starting procedure is to place a main microphone 'three metres up and three metres back' from the conductor's head, with a second microphone perhaps further back and a little higher. The outputs of the two microphones are then compared. The one giving the better overall blend without losing the detail then remains and the other microphone is moved to another place for a second comparison. An alteration in distance of even a few centimetres can make a significant difference to the sound produced. To avoid the effects of either the reinforcement or the phase cancellation of the reflected sound, microphones are best placed asymmetrically within a hall, off the centre line. Similarly one avoids the axis of a concave surface such as a dome.

The process of comparison continues until no further improvement can be obtained. It is essential that such listening is done with reference to the actual sounds produced by the musicians. It is important that producers and sound balance engineers do not confine themselves to the noises produced by their monitoring equipment, good though they may be, but listen in the hall to what the performers are doing.

Having achieved the best placing for the main microphone, soloists', 'filler' or 'space' mics may be added. These additional microphones must not be allowed to take over the balance nor should their use alter the 'perspective' during a concert. For a group that is balanced internally the only additional control is likely to be a compression of the dynamic range as the music proceeds. An intelligent anticipation of variations in the dynamics of the sound is required and although a limiter/compressor will avoid the worst excesses of overload, it can also iron out all artistic subtlety in the performance. There is no substitute for a manual control that takes account of the information from the musical score and is combined with a sensitive appreciation of what comes out of the loudspeaker.

The aim is to broadcast the music together with whatever atmosphere may be appropriate. To secure the confidence of the conductor, bandmaster or leader, it is good practice to invite him into the sound mixing and monitoring area to listen to the balance achieved during rehearsal. It should be remembered that since he normally stands close to the musicians, he is used to a fairly high sound level and the playback of rehearsal recordings should take this into account. Recordings should never be played back into the room in which they were made since this can lead to an undue emphasis of any acoustic peculiarity. A balance should also be monitored at low level, at what more closely resembles domestic listening conditions. This is particularly important with vocal work when the clarity of diction may suffer if the final balance is only judged at high level.

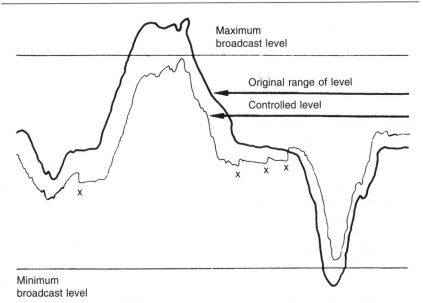

Maximum broadcast level

Original range of level

Controlled level

Minimum broadcast level

Figure 18.3 Dynamic compression. The range of loudness in the studio can easily rise above the maximum and fall below the minimum levels acceptable to the broadcast system. By anticipating a loud passage the main fader is used to reduce the level before it occurs. Similarly the level prior to a quiet part is increased. Faders generally work in small steps, each one introducing a barely perceptible change (X)

Creation of a synthetic balance

Whereas the reproduction of an existing sound calls for the integration of performance and acoustic using predominantly one microphone, the creation of a synthetic balance, which in reality exists nowhere, requires the use of many microphones to separate the musical elements in order to 'treat' and reassemble them in a new way. For example, a concert band arrangement calling for a flute solo against a backing from a full brass section would be impossible unless for the duration of the solo the flute can be specially favoured at the expense of the brass. Achieving this relies on the ability to separate the flute from everything else in such a way that it can be individually emphasised without affecting the sound from other instruments. The factors involved in achieving this separation are: studio layout, microphone types, source treatment, mixing technique, and recording technique.

Studio layout

The physical arrangement of a music group has to satisfy several criteria, some of which may conflict:

1 To achieve 'separation', quiet instruments and singers should not be too close to loud instruments.

2 The spatial arrangement must not inhibit any stereo effect required in the final balance.
3 The conductor or leader must be able to see everyone.
4 The musicians must be able to hear themselves and each other.
5 Certain musicians will need to see other musicians.

The producer should not force players to adopt an unusual layout against their will since the standard of performance is likely to suffer. He must resolve any difficulty by suggesting alternative means of meeting the musical requirements. For example, a rhythm section of piano, bass, drums and guitar needs to be tightly grouped together so that they can all see and hear each other. If there is a tendency for them to be picked up on the microphones of adjacent instruments they may need to be screened off. If this in turn affects their or other musicians' ability to see or hear, the sight-line can be restored by the use of screens with inset glass panels, and aural communication maintained by means of headphones – or a foldback speaker on the floor near them – fed with whatever mix of other sources they need. Working from a floor-plan may inhibit thinking in three dimensions. The solution to a sight-line problem may be to use rostra or staging to raise, say, a brass section – this can improve the separation by keeping it off other mics, which are pointing downwards.

Left to themselves, musicians often adopt the layout usual to them for

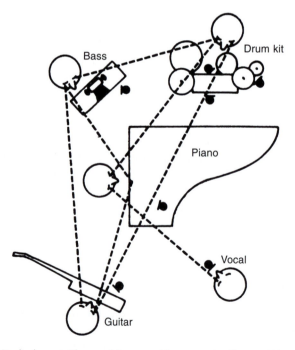

Figure 18.4 Studio layout. The musicians need to see each other and the microphones placed so as to avoid undue pick-up of other instruments

cabaret or stage work. This may be unsuitable for broadcasting and it is essential that the producer knows the instrumentation detail beforehand in order to make positive suggestions for the studio arrangements. He must know whether bass and guitar players have an electric or acoustic instrument. Also how much 'doubling' is anticipated, that is, one musician playing more than one instrument, sometimes within the same musical piece, and whether the players are also to sing. This information is also useful in knowing how many music stands to provide.

It is preferable to avoid undue movement in the studio during recording or transmission but it may be necessary to ask a particular musician to move to another microphone, for example for a solo. It is also usual to ask brass players to move into the microphone slightly when using a mute. This avoids opening the fader to such an extent that the separation might be affected.

Microphones for music

As an aid in achieving separation, the sound recordist's best friend is the directional microphone. The physical layout in the studio is partly arrived at in the light of what microphones are available. A ribbon microphone with its figure of eight directivity pattern is most useful for its lack of pick-up on the two sides. Used horizontally just above a flute or piano, it is effectively dead to other instruments in the same plane. A cardioid microphone gives an adequate response over its front 180° and will reject sounds arriving at the back – good for covering a string section. A hypercardioid microphone will narrow the angle of acceptance giving an even more useful area of rejection. Condenser or electret types of microphone with variable directivity patterns are valuable for the flexibility they afford in being adjustable – often by remote control – to meet particular needs.

The presenter of a live music show with an audience will probably need his or her mic to be fed to a public address system as well as to the broadcaster's mixer. Careless PA suffers from the all too familiar howl-round problem, but a personal radio-mic is a great help – provided that the sound mixer can see the stage. PA mixing is best done in the hall itself so that the operator hears what the live audience hears and can take immediate action in the event of the onset of a 'howl'. If this is not possible, a video link is extremely useful in allowing both the PA and the broadcast mixers to see what the performers are doing. The producer should also decide if the presenter needs talkback via an earphone.

The wide operatic platform can be covered by a suspended stereo pair, or a row of three, front-of-stage mics. The low 'boundary effect' mic is particularly effective in this context.

The closer a microphone is placed to an instrument, the greater the relative balance between it and the sounds from other sources. However, while separation is improved, other effects have to be considered:

1 The pick-up of sound very close to an instrument may be of inferior musical quality. It can be uneven across the frequency range or sound 'rough' due to the reproduction of harmonics not normally heard.
2 There may be an undue emphasis of finger noise or the instrument's mechanical action.
3 The volume of sound may produce overload distortion in the microphone, or in the subsequent electronics.
4 Movement of the player relative to the microphone becomes very critical, causing significant variations in both the quantity and quality of the sound.
5 Close microphone techniques require more microphones and more mixing channels thus increasing the complexity of the operation and the possibility of error.

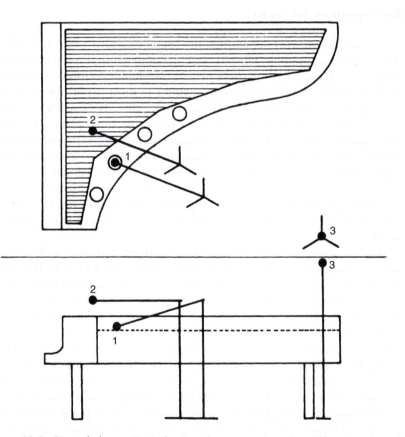

Figure 18.5 Piano balance. Typical microphone positions for different styles of music. 1. Emphasises percussive quality for pop music and jazz. 2. Broader intermediate position for light music. 3. Full piano sound for recitals and other serious or classical music

The choice and placing of microphones around individual instruments is a matter of skill and judgement by the recording engineer. But no matter how complex the technical operation, the producer must also be aware of other considerations. He must not allow the changing of microphones, alterations to the layout, the running of cables or audio feeds to interfere unduly with the music-making or human relations aspects. It is possible to be technically so pedantic as to inhibit the performance. If major changes in the studio arrangements are required, it is much better to do them while the musicians are given a break. Under these circumstances the broadcaster has an obvious additional responsibility to safeguard instruments left in the studio.

The ultimate in separation is to do away with the microphone altogether. This is possible with certain electric instruments where their output is available via an appropriately terminated lead, as well as acoustically for the benefit of the player and other musicians. The lead should be connected through a 'direct inject' box to a normal microphone input cable. Used particularly in conjunction with electric guitars, the box obviates the hum, rattles and resonances often associated with the alternative method, namely placing a microphone in front of the instrument's loudspeaker.

A further point in the use of any electrically powered instrument is that it should be connected to a studio power socket through a mains isolating transformer. This will protect the power supply in the event of a failure within the instrument and will exclude the possibility of an electrocution accident – so long as the studio equipment and the individual instrument are correctly wired and properly earthed. Faulty wiring or earthing arrangements can cause an accident, for example in the event of a performer touching simultaneously two 'earths' ('grounds') – his instrument and a microphone stand – which in fact are at a different potential. While comparatively rare it is a matter which needs attention in the broadcasting of amateur pop groups using their own equipment.

A number of treatments may be applied to individual sound sources, some of which can affect the separation. These include frequency control, dynamic control, and echo.

Frequency control[1]

The tone quality of any music source can be altered by the emphasis or suppression of a given portion of the frequency spectrum. Using a graphic equaliser or other frequency discriminating amplifier, a singer's voice is given added 'presence' and clarity of diction is improved by a lift in the frequency response in the octave between 2.8 kHz and 5.6 kHz. A string section can be 'thickened' and made 'warmer' by an increase in the lower and middle frequencies, while brass is given greater 'attack' and a sharper

---

[1] Sometimes referred to as EQ or equalisation. See Glossary.

edge by some 'top lift'. It should be noted, however, that in effectively making the microphone for the brass section more sensitive to the higher frequencies, spill-over from the cymbals is likely to increase and separation in this direction is therefore reduced. Fairly savage control is often applied to jazz or rhythm piano to increase its percussive quality. It is also useful on a one-microphone balance, particularly on an outside broadcast, to reduce any resonance or other acoustic effect inherent in the hall.

Dynamic control

This can be applied automatically by inserting a compressor/limiter device in the individual microphone chain. Once set, the level obtained from any given source will remain constant – quiet passages remain audible, loud parts do not overload. It becomes impossible for the flute to be swamped by the brass. Because of the economics of popular music, commercial recording companies have attained a high degree of sophistication in the use of dynamic control. This is unlikely to be reached by broadcasters who are not able to go to such lengths in recording their music. However, devices of this type can save studio time and their progressive application is likely. Variations include a 'voice-over' facility which enables one source to take precedence over another. Originally intended for DJs, its obvious use is in relation to singers and other vocalists but it can be applied to any source relative to any other. Computerised music desks can be programmed to store in a memory the changing mix required throughout a performance. In the final take the basics can be left to look after themselves.

Echo

The various means of adding the echo effect, more correctly described as reverberation, are the digital effects unit, the echo room, mechanical plate or spring, and tape echo. The *digital effects unit* creates any kind of echo electronically through a number of controls and presets which affect the duration, type, mix, and decay characteristic of the reverberation. It is also possible to alter the frequency response, thereby producing any acoustic from a boxy telephone booth to a cave or cathedral.

This unit is extremely versatile and useful for all music recording. It not only offers echo and delay, but facilities may also include dynamic compression, equalisation, and noise reduction – even of a specific frequency – making it especially useful in controlling the howl-round of PA.

An *echo room* is a bare room with a live acoustic, often in the basement of the studio building. It contains a loudspeaker fed with the appropriate source, and facing away from it a microphone, mono or stereo as required,

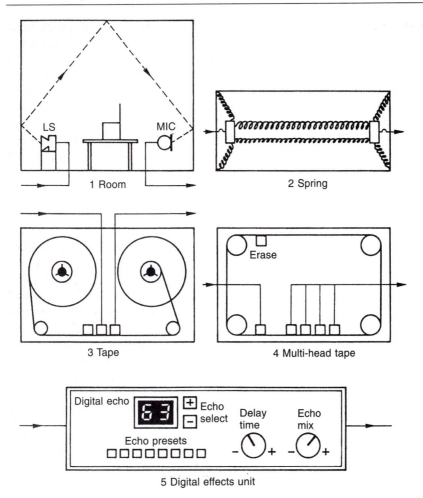

LS   MIC

1 Room   2 Spring

3 Tape   Erase   4 Multi-head tape

Digital echo   Echo
                    select   Delay   Echo
                                  time   mix
Echo presets
                                  – +   – +

5 Digital effects unit

Figure 18.6 Methods of creating echo. Echo device is fed with a programme signal which it delays. The reverberant output is returned to the mixing desk

which will pick up the loudspeaker output after its reflection from the wall surfaces. The output of the microphone is returned to the mixing desk as the echo source. An echo room should also contain several small obstructions – such as disused or broken furniture – to break up the natural standing wave patterns. In the absence of a permanent installation, an echo room may be rigged in a reverberant stair well or lavatory, so long as it is not in use!

The output from the source can be made to vibrate a *mechanical plate or spring*, the vibrations travelling through it to a transducer which converts them back into electrical energy, returning as echo to the mixing desk. Equivalent to a two-dimensional room, the reverberation time is adjustable depending on the mechanical damping applied. Some such devices are in portable form and have useful OB applications.

A crude form of echo, known as *tape echo*, is possible by feeding an output to a studio tape machine which records it; the playback head reproducing it a moment later acting as the echo source. The actual delay will depend on the tape speed and the distance between the record and replay heads. Recording machines designed expressly to produce reverberation effects have several replay heads with variable spacing to avoid the 'flutter' which can occur when all the sound is returned after a single constant delay.

When echo is added to solo singers' voices it is wise to let them listen on headphones since they will almost certainly want to adjust their phrasing to suit the new sound.

Mixing technique

Before mixing a multi-microphone balance, each channel should be checked for delivering its correct source, with adequate separation from its neighbours, having the desired amount of 'treatment', and producing a clear distortion-free sound. It is a sensible procedure to label the channel faders with the appropriate source information – solo vocal, piano, trumpets, and so on. For a stereo balance it is necessary to be clear about the placing of each instrument in the stereo picture. Stereo microphones are physically adjusted to spread their output across the required width of the picture, and the mono microphone 'pan-pots' set so that their placing coincides with that provided by any stereo microphone covering the same source. It is possible to create a stereo balance using only monophonic microphones by 'spotting' them across the stereo picture, but it is an advantage to have at least one genuine stereo channel, even if it is only the echo source.

Microphones should be mixed first in their 'family' groups. The rhythm section, strings or brass, is individually mixed to obtain an internal balance. The sections are then added to each other to obtain an overall mix. Large music desks allow their channels to be grouped together so that sections can be balanced relative to each other by the operation of their 'group' fader without disturbing the individual channel faders. Successful mixing requires a logical progression for if all the faders are opened to begin with, the resulting confusion can be such that it is very difficult to identify problems. It is important to arrive at a trial balance fairly quickly since the ear will rapidly become accustomed to almost anything. The overall control is then adjusted so that the maximum level does not exceed the permitted limit.

The prime requirement for a satisfying music balance is that there should be a proper relationship between melody, harmony and rhythm. Since the group is not internally balanced, listening on the studio floor will not indicate the required result. Only the person with his hands on the faders and listening to the loudspeaker is in a position to arrive at a final result. The melody will almost certainly be passed around the group, and this will need very precise manipulation of the faders. If the strings take the tune from an up-beat, their faders must be opened further *from that point* – no earlier otherwise the perspective will alter. At the end of that section the faders must be returned to their normal position to avoid loss of separation. It is probable that alterations to a source making use of echo will need a corresponding and simultaneous variation in the echo return channel. It is a considerable help therefore to have an elementary music score or 'lead sheet' which indicates which instrument has the melody at any one time. If this is not provided by the musical director, the producer should ensure that notes are taken at the rehearsal – for example, how many choruses there are of a song, at what point the trumpets put their mutes in, and perhaps more importantly, when they take them out again, when the singer's microphone has to be 'live' and so on. A professional music balance operator quickly develops a flair for reacting to the unexpected but, as with most aspects of broadcasting, some basic preparation is only sensible.

It is important that faders that are opened to accentuate a particular instrument are afterwards returned to their normal setting. Unless this is done it will be found that all the faders are gradually becoming further and further open with a compensating reduction of the overall control. This is clearly counter-productive and leads to a restriction in flexibility as the channels run out of 'headroom'.

As with all music balance work, the mixing must not take place only under conditions of high level monitoring. The purpose of having the loudspeaker turned well up is so that even small blemishes can be detected and corrected. But from time to time the listening level must be reduced to domestic proportions to check particularly the balance of vocals against the backing, and the acceptable level of applause. The ear has a logarithmic response to volume and is far from being equally sensitive to all the frequencies of the musical spectrum. This means that the perceived relationship between the loudness of different frequencies heard at high level is not the same as when heard at a lower level. Unless the loudspeaker is used with domestic listening in mind as well as for professional monitoring there will be a tendency as far as the listener at home is concerned to underbalance the echo, the bass, and the extreme top frequencies. For this reason many professional studios check the final mix on a simple domestic loudspeaker.

Recording technique

The successful handling of a music session requires the cooperation of everyone involved. There are several ways of actually getting the material recorded and the procedure to be adopted should be agreed at the outset.

The first method is obviously to treat a recording in the same way as a live transmission. One starts when the red light goes on and continues until it goes out. This will be the procedure at a public concert when, for example, no retakes are possible.

Second, a studio session with an audience present. Minor faults in the performance or mixing will be allowed to pass but the producer must decide when something has occurred that necessitates a retake. The audience may be totally unaware of the problem but he will have lightly and briefly to explain the existence of 'a gremlin' and detail quickly to the musical director the need to retake from point A to point B.

Third, without an audience the musicians can agree to use the time as thought best. This may be to rehearse all the material and then to record it, or to rehearse and record each individual item. In either case breaks can be taken as and when decided by the producer in consultation with the musical director. A music producer must of course be totally familiar with any agreements with the performers' unions which are likely to affect how a session is run.

Fourth, to record the most material in the shortest time or to accommodate special requirements, the music may be 'multi-tracked'. This means that instead of immediately arriving at a final mix, the individual instruments or groups of channels are recorded on perhaps 16 or 32 separate tracks of a 50 mm (2-inch) wide tape using an analogue recorder. A digital recorder can put down 48 tracks on 12.7 mm ($\frac{1}{2}$-inch) tape. However, such machines are expensive and the much cheaper alternative is to download tracks digitally to a computer equipped with multi-tracking software. This provides a wide range of signal manipulation, monitoring, auto-location, and editing. Since some tracks can be set to play while others record, the subsequent process of 'mixing down' or 'reduction' is relatively straightforward. This facility to multi-track enables musicians to record their contribution without the necessity of having all the performers present simultaneously. It also permits 'doubling', that is, the performing of more than one role in the same piece of music. This double-tracking is most often used for an instrumental group who also sing. The musicians first play the music – either conventionally recorded or multi-tracked – and when this is played back to them on headphones in the form of a 'sync output' actually generated by the record heads they add the vocal parts. These are recorded on separate tracks or mixed with the music tracks to form a final recording. The process can be repeated to enable singers to sing with themselves as many times as is required to create the desired sound. An added advantage of multi-tracking is that any mistake does not spoil the whole performance – a retake may only by required to correct the individual error.

Since recording at different times represents total separation, multi-tracking techniques can solve other problems. For instance the physical separation or screening of loud instruments such as a drum kit can be avoided by recording the rhythm section separately – usually first, in order to keep subsequent performers in sync. It is an especially useful technique to apply to music recording which has to be done in a small non-purpose-built studio, and even a four-track recorder can help to produce a more professional sound.

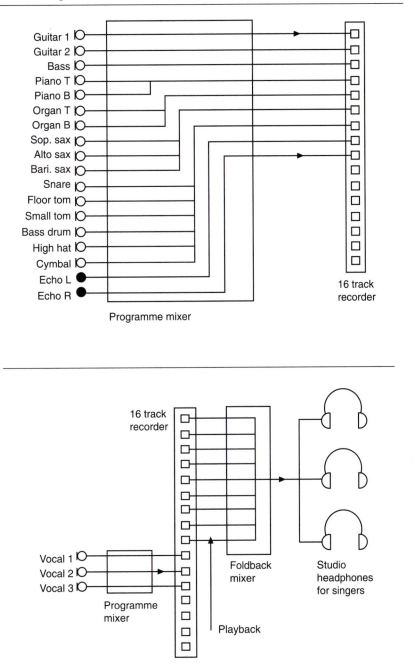

Figure 18.7 Multi-track recording. In this example the 18 original sources are first recorded as nine separate tracks. A further three vocal tracks are then added, the singers listening to the output of the record heads to maintain synchronisation. The 12 tracks are subsequently mixed together in a reduction session to make the final recording

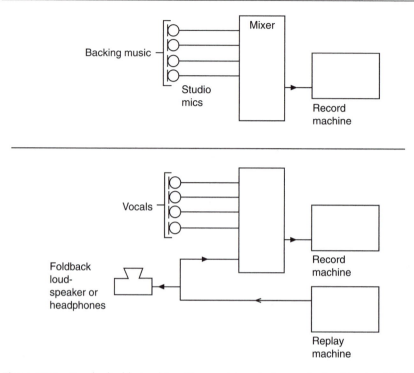

Figure 18.8 Simple double tracking. The musicians first record a 'backing track'. This is then replayed so that it is heard on the studio foldback loudspeaker (or headphones) while at the same time mixed with the output of the studio microphones. In this example the final recording consists of eight performances, although of course the process can be repeated further

Unless special noise-reduction facilities or DAT recording is employed, the re-recording of ordinary analogue tape tracks will increase the hiss level on the final recording. It is essential therefore that the best equipment is used and the original recordings are made at the highest tape speed available. A point worth noting by studios unable to afford the latest in sophisticated stereo recording equipment is that a good domestic video recorder connected to a digital processor is capable of excellent results.

Production points

Essentially the producer's job is to obtain a satisfactory end product within the time available. He should avoid running into overtime which with professional musicians is expensive. Neither should he be an oppressively hard taskmaster so that the session ends with time to spare, and everyone is relieved to get out. The producer must remember that all artists need encouragement and he has a positive duty to contribute to their giving the best performance possible.

He avoids using the general talkback into the studio to make disparaging comments about individual performers. Instead, he directs all his remarks to the conductor or group leader via a special headphone feed, or better still by going into the studio for a personal talk.

He fosters a professionally friendly atmosphere and sounds relaxed even when under pressure. He is often the mediator between the performers and the technical staff, he explains delays and agrees new timescales. He anticipates the need for breaks and rest periods. During these times it is often sensible to invite the MD or group leader into the control cubicle to hear the recorded balance and discuss any problems.

He avoids displaying a musical knowledge which, while attempting to impress, is based on uncertain foundations. 'Can we have a shade more "arco" on the trombones?' is a guaranteed credibility loser.

He makes careful rehearsal notes for the recording or transmission, and takes details of the final timings and any retakes for subsequent editing.

When an audience is present he decides how it is best handled and what kind of introduction and warm-up is appropriate. He may have to do this himself.

He resolves the various conflicts which arise – whether the air conditioning or heating is working properly, whether the lights are too bright or whether the banging in another part of the building can be stopped. He remembers other needs – ashtrays, drinking water, lavatories, telephones.

When a retake is required, he makes a positive decision and communicates it quickly to the musical director. The producer exercises the broadcaster's normal editorial judgement and is responsible for the quality of the final programme.

He ensures that the musicians are paid – those performers who signed a contract and made the music and, through the appropriate agencies, those who wrote, arranged and published the work still in copyright.

After the session he says thank you – to everyone.

19

Drama – principles

The radio medium has a long and distinguished history of turning thoughts, words and actions into satisfying pictures within the listener's mind by using the techniques of drama. But there is no need for the producer to think only in terms of the Shakespeare play – the principles of radio drama apply to the well-made commercial, a programme trail, dramatised reading, five-minute serial or two-minute teaching point in a programme for schools. Since the size and scope of the pictures created are limited only by the minds which devise and interpret them, the medium in its relationship to drama is unique, and any radio service is the poorer for not attempting to work in this area.

As an illustration of the effective simplicity of the use of sound alone, listen now to the celebrated example by Stan Freberg who gave it as part of his argument for selling advertising time on radio.

MAN: Radio? Why should I advertise on radio? There's nothing to look at . . . no pictures.

GUY: Listen, you can do things on radio you couldn't possibly do on TV.

MAN: That'll be the day.

GUY: Ah huh. All right, watch this. (Clears throat). OK, people, now when I give you the cue, I want the 700-foot mountain of whipped cream to roll into Lake Michigan which has been drained and filled with hot chocolate. Then the Royal Canadian Air Force will fly overhead towing the 10-ton maraschino cherry which will be dropped into the whipped cream, to the cheering of 25,000 extras. All right . . . cue the mountain . . .

SOUND: Groaning and creaking of mountain into big splash.

GUY: Cue the Air Force!

SOUND: Drone of many planes.

GUY: Cue the maraschino cherry . . .

SOUND: Whistle of bomb into bloop! of cherry hitting whipped cream.

GUY: Okay, twenty-five thousand cheering extras . . .

SOUND: Roar of mighty crowd. Sound builds up and cuts off sharp!

```
GUY:        Now . . . you wanta try that on television?

MAN:        Well . . .

GUY:        You see . . . radio is a very special medium, because it
            stretches the imagination.

MAN:        Doesn't television stretch the imagination?

GUY:        Up to 21 inches, yes.
```

[Courtesy Freberg Ltd]

A story can offer a framework for the understanding – or at least an interpretation – of life's events. Often a mirror in which we can see ourselves – our actions, motives, and faults – and the outcomes and results can contribute to our own learning. Drama is about conflict and resolution, relationships and feelings and people being motivated by them, both driving and driven by events. What happens should be credible, the people believable, and the ending have a sense of logic however unusual and curious so that the listener does not feel cheated or let down.

The aim with all dramatic writing is for the original ideas to be re-created in the listener's mind and since the end result occurs purely within the imagination, there are few limitations of size, reality, place, mood, time or speed of transition. Unlike the visual arts where the scenery is provided directly, the listener to radio supplies his own mental images in response to the information he is given. If the 'signposts' are too few or of the wrong kind, the listener becomes disorientated and cannot follow what is happening. If there are too many the result is likely to be obvious and 'corny'. Neither will satisfy. The writer must therefore be especially sensitive to how his audience is likely to react – and since the individual images may stem largely from personal experience, of which the writer of course knows nothing, this is not easy. But it is the ageless art of the storyteller – saying enough to allow listeners to follow the thread but not so much that they do not want to know what is to happen next or cannot make their own contribution.

The writer must have a thorough understanding of the medium and the production process, while the producer needs a firm grasp of the writing requirements. If they are not one and the same person, there must be a strong collaboration. There can be no isolation, but if there is to be a dividing line, let the writer put everything down knowing how it is to sound, while the producer turns this into the reality of a broadcast knowing how it is to be 'seen' in the mind's eye.

The component parts with which both are working are speech, music, sound effects, and silence.

The idea

Before committing anything to paper, it is essential to think through the basic ideas of plot and form – once these are decided, a great deal follows naturally. The first question is to do with the material's suitability for the target audience, the second with its technical feasibility.

Assuming that the writer is starting from scratch and not adapting an existing play, he should be clear as to his broad intention – to make people laugh, to comment on or explain a contemporary situation, to convey a message, to tell a story, to entertain. How can he best enable the listener to 'connect' with his intention? Does he want him to identify with one of the characters? Should the basic situation be one with which the listener can easily relate?

The second point at this initial stage is to know whether the play has to be written within certain technical or cost limitations. To do something simple and well is preferable to failing with something complicated. There seems little point in writing a play which calls for six sound effects turntables or CD players, echo, a variety of acoustics, distorted voice-over, and a crowd chanting specific lines of script, if the studio facilities are not able to meet these demands. Of course with ingenuity even a simple studio can provide most if not all these devices. But the most crucial factor is often simply a shortage of time. There may be limitations too in the capabilities of the talent available. The writer, for example, should beware creating a part which is emotionally exceptionally demanding only to hear it inadequately performed. Writing for the amateur or child actor can be very rewarding in the surprises which their creative flowering may bring, but it can also be frustrating if you automatically transfer into the script the demands and standards of the professional stage.

Thus the writer must know at the outset how to tailor his play for the medium, what he is attempting to put over, how he expects his audience to relate to the material, and whether what he wants is technically and financially possible. From here there are three possible starting points – the story, the setting or the characters.

Story construction

The simplest way of telling a story is to:

1 Explain the situation.
2 Introduce 'conflict'.
3 Develop the action.
4 Resolve the conflict.

Of course there may be complications and sub-plots but the essence of a good story is to want to find out 'what happens in the end'. Who committed the crime? Were the lovers reunited? Did the cavalry arrive in time? The element which tends to interest us most is the resolution of conflict and since this comes towards the end, there should be no problem

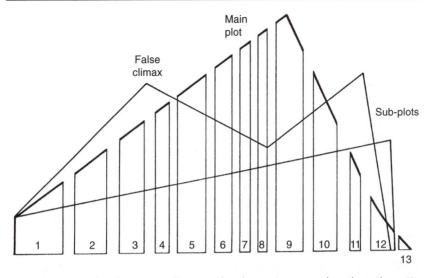

Figure 19.1 A plot diagram to illustrate the shortening scene length as the action rises to its climax:

1–2   introduction, setting and context, characterisation established;
3–5   conflict, events resulting from characters in the situation;
6–8   rising action, complication, suspense;
9   maximum tension, crisis, climax;
10–11   falling action, resolution;
12–13   denouement, twist.

of maintaining interest once into the 'rising action' of the play. And in the final scene it is not necessary to tie up all the loose ends – to dot every 'i' and cross every 't' in a neat and tidy conclusion. Life seldom works that way. It is often better to have something unsaid, leaving the listener still with a question, an issue, or a motivation to think about. Parables are stories which deliberately do not go straight from A to B, but take a parabolic route leaving the hearer to work out the implications of it all. This is one of the fascinations of the story form.

In radio, scenes can be shorter than in the theatre, and intercutting between different situations is a simple matter of keeping the listener informed about where he is at any one time. This ability to move quickly in terms of location should be used positively to achieve a variety and contrast which itself adds interest. The impact of a scene involving a group of fear-stricken people faced with impending disaster is heightened by its direct juxtaposition with another, but related, group unaware of the danger. If the rate of intercutting becomes progressively faster and the scenes shorter, the pace of the play increases. This sense of acceleration, or at least of movement, may be in the plot itself but a writer can inject greater excitement or tension simply in his handling of the scene length and in their relation to one another. Thus the overall shape of the play may be a steady development of its progress, heightening and increasing.

Or it may revolve around the stop-go tempo of successive components. Interest through contrast may be obtained by a variety of means, for example:

1 Change of pace: fast/slow action, noisy/quiet locations, long/short scenes.
2 Change of mood: tense/relaxed atmosphere, angry/happy, tragic/dispassionate.
3 Change of place: indoors/open air, crowded/deserted, opulent/poverty stricken.

The radio writer is concerned with images created by sound alone; if he wants colour and mood he must paint them with the words he uses and choose locations that are aurally evocative.

On a personal note to illustrate the impact of contrast within a play, I recall a moment from a drama on the life of Christ. The violence and anger of the crowd demanding his execution is progressively increased, the shouting grows more vehement. Then we hear the Roman soldiers, the hammering of nails and the agony of the crucifixion. Human clamour gives way to a deeper, darker sense of tragedy and doom. Christ's last words are uttered, a crash of thunder through to a climax of discordant music gradually subsiding, quietening to silence. A pause. Then slowly – birdsong.

How wrong it would have been to spoil that contrast by using the narrator, as elsewhere in the play.

The setting

Situation comedy drama often begins with a setting which then becomes animated with characters. The storyline comes later driven by the circumstances, generally a series of predicaments in which the characters find themselves. Fortunately radio can provide almost any setting at will – a royal household in ancient Egypt, a space capsule journeying to a distant planet, or a ranch house in outback Australia. The setting, plus one or two of the principal characters, may be the key theme holding a series of plays – or advertisements – together. It is important that the time and place is well researched so that credibility is maintained for those listeners who are expert in the particular situation.

Within a play the setting can obviously vary considerably and one of the devices used to create interest is to have a strong contrast of locale in adjacent scenes, for example to move from an opulent modern office peopled by senior managers to a struggling rural hospital affected by their decisions. Changes of location are very effective when run in parallel with changes in disposition or mood.

Characterisation

One of Britain's best-known advisers in this area, Bart Gavigan – often referred to as a 'script doctor' – says that there are three questions to be answered for a compelling story:

Who is the hero – or heroine?
What does he, or she, want?
Why should I care?

This last question emphasises the need for at least main characters to develop a rapport with the listener. They must be more than thinly described cardboard cutouts, they should appear as credible people with whom I can identify and whose cause I can care about. The writer has to create a play – or 30 second advertisement – so that the human motivation and behaviour is familiar. This does not mean that the setting has to be the same as the domestic circumstances of the audience – far from it. For whatever the setting the listener will meet people he can recognise – fallible, courageous, argumentative, greedy, fearful, compassionate, lazy, and so on. Characterisation is a key ingredient and it is important to sketch out a pen portrait of each character. This helps to stabilise them as people and it is easier to give them convincing dialogue. Here are some headings:

- their age, sex, where they live and how they talk
- height, weight, colouring, and general appearance
- their social values, sense of status, beliefs
- the car they drive, the clothes they like, how much money they have
- their family connection, friends – and enemies
- the jokes they make – or not, how trusting they are, how perceptive
- their moods, orientations, preferences and dislikes.

Characters have faults, they fall apart in crises; they will appear illogical because their words and their deeds won't always coincide. They will reveal themselves in what they say, but even more in what they do. One of the necessary tensions in compelling drama is the inner conflict that exists in human beings – the inconsistencies between what I want to do, what I ought to do, and what I actually do. Saints will have their failings and even the worst sinners may have their redeeming features under certain circumstances. Substantial characters convey real human complexity and the writer cannot accurately portray them until he knows them. When characterisation is fully in place the writer may let the characters almost write themselves since they know how they are likely to react to a set of events driven by the story. Writers and producers should tell actors as much as they can about the characters – their personality, typical behaviour, and disposition.

Dialogue

'Look out, he's got a gun.' Lines like this, unnecessary in film, television or theatre where the audience can see that he has a gun, are essential in radio as a means of conveying information. The difficulty is that such 'point' lines can so easily sound contrived and false. All speech must be the natural colloquial talk of the character by whom it is uttered. In reproducing a contemporary situation, a writer can do no better than to take his notebook to the marketplace, restaurant or party and observe what people actually say, and their manner of saying it. Listen carefully to the talk of shoppers, eavesdrop on the conversation in the train. It is the stuff of your reality. And amid the talk, silence can be used to heighten a sense of tension or expectation. Overlapping voices convey anger, passion, excitement or crisis. Not only do radio characters say what they are doing, but people reveal their inner thoughts by helpfully thinking aloud, or saying their words as they write a letter. These are devices for the medium and should be used with subtlety if the result is to feel true to life.

Producers should beware writers who preface a scene with stage directions: 'The scene is set in a lonely castle in the Scottish Highlands. A fire is roaring in the grate. Outside a storm is brewing. The Laird and his visitor enter.'

Such 'picture setting' designed for the reader rather than the listener should be crossed out and the dialogue considered in isolation. If the words themselves create the same scene, the directions are superfluous; if not, the dialogue is faulty:

```
Laird:     (Approaching) Come in, come in. It looks like a storm is on the
           way.
Visitor:   (Approaching) Thank you. I'm afraid you're right. It was starting
           to rain on the last few miles of my journey.
Laird:     Well come and warm yourself by the fire; we don't get many
           visitors.
Visitor:   You are a bit off the beaten track, but since I was in the
           Highlands and've always been fascinated by castles I thought I
           would call in - I hope you don't mind.
Laird:     Not at all, I get a bit lonely by myself.
Visitor:   (Rubbing hands) Ah, that's better. This is a lovely room - is
           this oak panelling as old as it looks?
                          etc. etc.
```

The producer is able to add considerably to such a scene in its casting, in the voices used – for example, in the age and accents of the two characters, and whether the mood is jovial or sinister.

In addition to visual information, character and plot, the dialogue must remind us from time to time of who is speaking to whom. Anyone 'present' must either be given a line or be referred to, so that they can be included in the listener's mental picture:

```
Andrew:   Look, John, I know I said I wouldn't mention it but . . . well
          something's happened I think you should know about.
John:     What is it, Andrew? What's happened?
```

The use of names within the dialogue is particularly important at the beginning of the scene.

Characters should also refer to the situation not within the immediate picture so that the listener's imagination is equipped with all the relevant information:

```
Robins:   . . . And how precisely do you propose to escape? There are
          guards right outside the door, and more at the entrance of the
          block. Even if we got outside, there's a barbed wire fence two
          storeys high – and it's all patrolled by dogs. I've heard them.
          I tell you there isn't a chance.
Jones:    But aren't you forgetting something – something rather obvious?
```

An obvious point which the writer does not forget is that radio is not only blind, but unless the drama is in stereo, it is half-deaf as well. Movement and distance have to be indicated, either in the acoustic or other production technique, or in the dialogue. Here are three examples:

```
(off mic) I think I've found it, come over here.
Look, there they are – down there on the beach. They must be half a mile
away by now.
(softly) I've often thought of it like this – really close to you.
```

We shall return later to the question of creating the effects of perspective in the studio.

To achieve a flow to the play, consecutive scenes can be made to link one into the other. The dialogue at the end of one scene points forward to the next:

```
Voice 1:  Well, I'll see you on Friday – and remember to bring the stuff
          with you.
Voice 2:  Don't worry, I'll be there.
Voice 1:  Down by the river then, at 8 o'clock – and mind you're not late.
```

If this is then followed by the sound of water, we can assume that the action has moved forward to the Friday and that we are down by the river. The actual scene-change is most often through a fade-out of the last line – a line incidentally like the second half of the one in this last example, words we can afford to miss. There is a moment's pause, and a fade-up on the first line of the new scene. Other methods are by direct cutting without fades, or possibly through a music link. The use of a narrator will almost always overcome difficulties of transition so long as the script avoids clichés of the 'meanwhile, back at the ranch' type.

A narrator is particularly useful in explaining a large amount of

background information which might be unduly tedious in conversational
form or where considerable compressions have to be made, for example
in adapting a book as a radio play. In these circumstances the narrator can
be used to help preserve the style and flavour of the original, especially
in those parts which have a good deal of exposition and description but
little action:

```
Narrator:   Shortly afterwards, Betty died and John, now destitute and
            friendless, was forced to beg on the streets in order to keep
            himself from starvation. Then one afternoon, ragged and nearing
            desperation, he was recognised by an old friend.
```

When in doubt, the experienced writer will almost certainly follow the
simplest course remembering that the listener will appreciate most what
he can readily understand.

Script layout

Following the normal standard of scripts intended for broadcast use, the
page should be typed on one side only to minimise handling noise, the
paper being of a firm 'non-rustle' type. The lines should be triple spaced
to allow room for alterations and actors' notes, and each speech numbered
for easy reference. Directions, or details of sound effects and music,
should be bracketed, underlined, or in capitals so that they stand out
clearly from the dialogue. The reproduction of scripts should be absolutely
clear and there should be plenty of copies so that spares are available.
      An example of page layout is shown below.

```
1.   _____   (INDOOR ACOUSTIC)

2.   BRADY:                 Why isn't Harris here yet? - You people at the
                            Foreign Office seem to think everyone has got
                            time to waste.

3.   SALMON:                I don't know, Colonel, it's not like him to be
                            late.

4.   BRADY:                 Well it's damned inconsiderate, I've a good mind
                            to . . .

5.   FX:     _____ KNOCK AT DOOR

6.   SALMON:                (RELIEVED) That's probably him. (GOING OFF)
                            I'll let him in.

7.   FX:     _____ DOOR OPENS OFF

8.   HARRIS:                (OFF) Hello, John.

9.   SALMON:                (OFF) Thank goodness you've arrived. We've been
                            waiting some time. (APPROACH) Colonel Brady I
                            don't think you've met Nigel Harris. He's our
                            representative . . .

10.  BRADY:                 (INTERRUPTING) I know perfectly well who he is,
                            what I want to know is where he's been.
```

11.	HARRIS:	Well, I've been trying to get us out of trouble. It's bad news I'm afraid. The money we had ready for the deal has gone, and Holden has disappeared.
12.	BRADY:	This is preposterous! Are you suggesting he's taken it?
13.	HARRIS:	I'm not suggesting anything, Colonel, but we know that last night he was at Victoria Station, – and bought a ticket for Marseilles.
14.	SALMON:	Marseilles? By train?
15.	HARRIS:	By train. At this moment I should think he, and the money, are half-way across Europe.
16.	FX:	TRAIN WHISTLE APPROACH AND ROAR OF TRAIN PASSING.
17.	FX:	CROSSFADE TO INTERIOR TRAIN RHYTHM CONTINUES UNDERNEATH.
18.	STEWARD:	(APPROACHING) Take your last sitting for lunch please. (CLOSER) Last sitting for lunch. Merci, Madame.

1.	FX:	COMPARTMENT DOOR SLIDES OPEN
2.	STEWARD:	(CLOSE) Last sitting for lunch, Sir. Excusez moi, Monsieur, – will you be wanting lunch? Monsieur? (TO SELF) C'est formidable. Quite a sleeper. (LOUDER) Excuse me, Sir, – Allow me to remove the newspaper.
3.	FX:	PAPER RUSTLE
4.	STEWARD:	Will you be . . . (GASP) Oh . . . Terrible . . . Terrible.
5.	FX:	TRAIN NOISE PASSING AND FADES INTO DISTANCE.
6.	FX:	PHONE RINGS. RECEIVER PICKED UP. RINGING STOPS.
7.	BRADY:	Hello.
8.	VOICE:	(DISTORT) Is that Colonel Brady?
9.	BRADY:	Yes. Who's that?
10.	VOICE:	(DISTORT) It doesn't matter but I thought you ought to know he's dead.
11.	BRADY:	Who's dead? Who is this?
12.	VOICE:	(DISTORT) Oh you know who's dead all right, – and I've got the money.
13.	BRADY:	You've got what money? Who are you?
14.	VOICE:	You'll know soon enough. I'll be in touch. . .
15.	FX:	TELEPHONE CLICK. DIALLING TONE.
16.	BRADY:	HELLO, hello . . . oh blast it.
17.	FX:	RECEIVER SLAMMED DOWN.
18.	FX:	OFFICE INTERCOM BUZZER.
19.	SECRETARY:	(DISTORT) Yes, Sir?

```
20.   BRADY:              Joan, I want you to get hold of Salmon and
                          Harris – can you do that?
21.   SECRETARY:          (DISTORT) Yes, Sir – they went back to the
                          Foreign Office.
22.   BRADY:              Well I want them – at once. And get me on
                          tonight's plane, – to Marseilles.
23.   CD:                 MUSIC TO END.
```

The actors

Casting a radio drama, whether it is a one-hour play or a short illustration, will nearly always end by being a compromise between who is suitable and who is available. Naturally, the producer will want the best performers but this is not always possible within the constraints of money. It is also difficult to assemble an ideal cast at one place, and possibly several times, for rehearsal and recording, and this to coincide with the availability of studio space. Again, it may be that two excellent players are available, but their voices are too similar to be used in the same piece. So there are several factors which will determine the final cast.

Actors new to radio have to recognise the limitations of the printed page, which is designed to place words in clearly readable lines. It cannot overlay words in the same way that voices can, and do:

```
Voice one:    The cost of this project is going to be 3 or 4 million –
              and that's big money by anyone's reckoning.
Voice two:    But that's rubbish, why I could do the job for . . .
Voice one:    (Interrupting) Don't tell me it's rubbish, anyway that's the
              figure and it's going ahead.
```

Although this is what might appear on the page, voice two is clearly going to react to the cost of the project immediately on hearing the figures – half-way through voice one's first line. The script writer may insert at that point (react) or (intake of breath) but generally it is best left to the imagination of the actor. Actors sometimes need to be persuaded to act, and not to become too script-bound. Then voice one interrupts. This does not mean that he waits until voice two has finished his previous line before starting with the 'Don't tell me . . .' He starts well before voice two breaks off, say on the word 'could'. The two voices will overlay each other for a few words thereby sounding more natural. Real conversation does this all the time.

On the matter of voice projection the normal speaking range over a conversational distance will suffice – intimate and confidential, to angry and hysterical. As the apparent distance is increased, so the projection also increases. In the following example, the actor goes over to the door and ends the line with more projection than he was using at the start. The voice gradually rises in pitch throughout the speech.

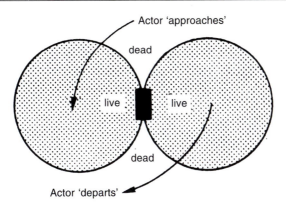

Figure 19.2 Movement on mic. The shaded circles are the normal areas of pick-up on both sides of a bi-directional microphone. An actor 'approaching' says his lines as he moves from the dead side to the live side, reducing his voice projection as he does so. Effective for indoor scenes

```
Voice:      Well I must go. (DEPARTS) I shan't be long but there are
            several things I must do. (DOOR OPENS, OFF) I'll be back as
            soon as I can. Goodbye. (DOOR SHUTS).
```

In moving to the 'dead' side of the microphone the actor's actual movement may have been no more than a metre. The aural impression given may be a retreat of at least five metres. It is important that such 'moves' are made only during spoken dialogue otherwise the actor will appear to have 'jumped' from a near to a distant position. Of course moving off-mic, which increases the ratio of reflected to direct sound, can only serve to give an impression of distance in an interior scene.

When the setting is in the open air, there is no reflected sound and distance has to be achieved by a combination of the actor's higher voice projection and the small volume derived from a low setting of the microphone channel fader. By this means it is possible to have a character shouting to us from 'over there', having a conversation with another person 'in the foreground' who is shouting back. Such a scene requires considerable manipulation of the microphone fader with no overlay of the voices. A preferred alternative is to have the actors in separate rooms, each with his own microphone and with a headphone feed of the mixed output.

The acoustic

In any discussion of monophonic perspective, distance is a function which separates characters in the sense of being near or far. The producer must always know where the listener is placed relative to the overall picture. Generally, but not necessarily, 'with' the microphone, the listener placed in a busily dynamic scene will need some information which distinguishes his following the action by moving through the scene, from his simply watching it from a static position.

Part of this distinction may be in the use of an acoustic which itself changes. Accompanied by the appropriate sound effects a move out of a reverberant courtroom into a small ante-room, or from the street into a telephone box, can be highly effective. There are four basic acoustics comprising the combinations of the quantity, and duration of the reflected sound:

1	No reverberation	Outdoor	Created by fully absorbent 'dead' studio
2	Little reverberation but long reverberation time	Library or large well-furnished room	Bright acoustic or a little reverberation added to normal studio
3	Much reverberation but short reverberation time	Telephone box, bathroom	Small enclosed space of reflective surfaces – 'boxy' acoustic
4	Much reverberation and long reverberation time	Cave, 'Royal Palace', concert hall	Artificial echo added to normal studio output

Frequency discrimination applied to the output of the echo device will add the distinctive coloration of a particular acoustic. The characteristic of a normal drama studio with a 'neutral' acoustic would limit the reverberation time to about 0.2 second. Associated with this would be a 'bright' area with a reverberation time of, say, 0.6 second, and a separate cubicle for a narrator. The key factor is flexibility so that by using screens, curtains and carpet, a variety of acoustic environments may be produced.

The use of tie-clip mics for actors avoids any pick-up of studio acoustic and allows considerable freedom of movement. Crossfading between a studio mic and a tie-clip mic with a suitable drop in vocal projection provides a very credible 'think piece' in the middle of dialogue.

Sound effects

When the curtain rises on a theatre stage the scenery is immediately obvious and the audience is given all the contextual information it requires for the play to start. So it is with radio, except that to achieve an unambiguous impact the sounds must be refined and simplified to those few which really carry the message. The equivalent of the theatre's 'backdrop' are those sounds which run throughout a scene – for example, rain, conversation at a party, traffic noise or the sounds of battle. These

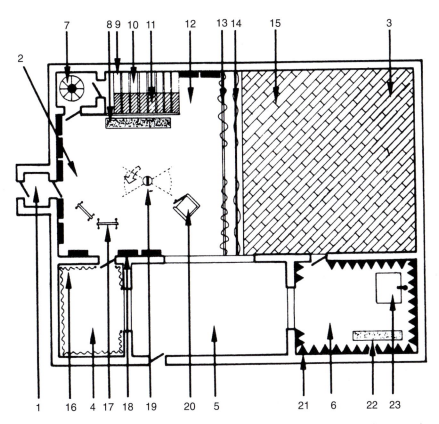

Figure 19.3 Features of a drama studio. **1** Entrance through sound lobby. **2** Studio 'dead' end. **3** Studio 'live' end. **4** Narrator's studio. **5.** Control cubicle with windows to all other areas. **6** Dead room – outdoor acoustic. **7** Spiral staircase – iron. **8** Gravel trough. **9** Sound effects staircase. **10** Cement tread stairs. **11** Wooden tread stairs. **12** Carpet floor. **13** Soft curtain drapes. **14** 'Hard' curtain – canvas or plastic. **15** Wood block floor. **16** Curtains. **17** Movable acoustic screen. **18** Acoustic absorber – wall box. **19** Bi-directional microphone. **20** Sound effects door. **21** Acoustic wedges – highly absorbent surface. **22** Sand or gravel trough. **23** Water tank

are most likely to be pre-recorded and reproduced from records, compact disc, cart or tape. The 'incidental furniture' and 'props' are those effects which are specially placed to suit the action – for instance, dialling a telephone, pouring a drink, closing a door or firing a gun. Such sounds are best made in the studio at the time of the appropriate dialogue, if possible by the actors themselves – for example, lighting a cigarette or taking a drink – but by someone else if hands are not free due to their holding a script.

The temptation for a producer new to drama is to use too many effects. While it is true that in the real world the sounds we hear are many and complex, radio drama in this respect purveys not what is real but what is understandable. It is possible to record genuine sounds which divorced from their visual reality convey nothing at all. The sound of a modern car drawing up has very little impact, yet it may be required to carry the dramatic turning point of the play. In the search for clear associations between situation and sound, radio over the years has developed conventions with generally understood meanings. The urgently stopping car virtually demands a screech of tyres, slamming doors and running footsteps. It becomes a little larger than life. Overdone, and it becomes comical.

Some other sounds which have become immediately understood are:

1   Passage of time – clock ticking.
2   Night time – owl hooting.
3   On the coast – seagulls and seawash.
4   On board sailing ship – creaking of ropes.
5   Early morning – cock crowing.
6   Out of doors, rural – birdsong.

The convention for normal movement is to do without footsteps. These are only used to underline a specific dramatic point.

Background sounds may or may not be audible to the actors in the studio, depending on the technical facilities available. It is important, however, for actors to know what they are up against, and any background sounds should be played over to them to help them visualise the scene and judge their level of projection. This is particularly important if the sounds are noisy – the cockpit of a light aircraft, a fairground or battle. If an actor has to react to a sound reproduced from a record or tape he will require a feed of the output either to his headphones or through a loudspeaker. The normal studio speaker is of course cut when any mic channel is opened, but an additional loudspeaker fed from the grams, CD and tape sources can remain on, unaffected by this muting. This facility is called *foldback*, and has the advantage not only that the cast can hear the effects, even though their own mics are live, but that any sounds reproduced in this way and picked up on the studio microphones will share the same acoustic as the actors' voices. Producers working in drama will soon establish their own methods of manufacturing studio sounds. The following are some which have saved endless time and trouble:

1   Walking through undergrowth or jungle – a bundle of recording tape rustled in the hands.
2   Walking through snow – a roll of cotton wool squeezed and twisted in the hands, or two blocks of salt rubbed together.
3   Horses' hooves – halved coconut shells are still the best from pawing the ground to a full gallop. They take practice though. A bunch of keys will produce the jingle of harness.
4   Pouring a drink – put a little water into the glass first so that the sound starts immediately the pouring begins.

5   Opening champagne – any good sound assistant ought to be able to make a convincing 'pop' with their mouth, otherwise blow a cork from a sawn-off bicycle pump. A little water poured on to Alka-Seltzer tablets or fruit salts close on-mic should do the rest.

6   A building on fire – cellophane from a cigarette packet rustled on-mic plus the breaking of small sticks.

7   Marching troops – a marching box is simply a cardboard box approximately $20 \times 10 \times 5$ cm, containing some small gravel. Held between the hands and shaken with precision it can execute drill movements to order.

8   Creaks – rusty bolts, chains or other hardware are worth saving for the appropriate aural occasion. A little resin put on a cloth and pulled tightly along a piece of string fastened to a resonator is worth trying.

9   In the case of a costume drama some silk or taffeta material rustled near the mic occasionally is a great help in suggesting movement.

The essential characteristic for all electronic or acoustic sound-making devices is that they should be simple, reliable and consistent. And if the precise effect cannot be achieved, it is worth remembering that by recording a sound and playing it back at a different speed, it can be altered in 'size' or made unrelatable to the known world. Hence the fantasy sounds varying from dinosaurs to outer space. Voices too can be made 'unhuman' by the use of a digital effects unit or 'harmoniser'. Unlike the recorded tape method a voice can be made lighter or deeper without affecting its speed of delivery. Time spent exploring the possibilities of digital effects will pay dividends for the drama producer.

In making a plea for authenticity and accuracy, it is worth noting that such attention to detail saves considerable letter-writing to those among the audience who are only too ready to display their knowledge. Someone will know that the firing of the type of shot used in the American War of Independence had an altogether characteristic sound, of course certain planes used in the Australian Flying Doctor services had three engines not two, and whoever heard of an English cuckoo in February?! The producer must either avoid being too explicit, or he must be right.

Music

An ally to the resourceful producer, music can add greatly to the radio play. However, if it is over-used or badly chosen, it becomes only an irritating distraction. The producer must decide in which of its various roles music is to be used:

1   As a 'leitmotif' to create an *overall style*. Opening and closing music plus its use within the play as links between some of the scenes will provide thematic continuity. The extracts are likely to be the same piece of music, or different passages from the same work, throughout.

2   Music chosen simply to *create mood* and establish the atmosphere of
     a scene. Whether it is 'haunted house' music or 'a day at the races',
     music should be chosen that is not so familiar that it arouses in the
     listener his own preconceived ideas and associations. In this respect
     it pays the producer to cultivate an awareness of the lesser known
     works in his library.
3   Reiterative or relentless music can be used to mark the *passage of
     time*, thus heightening the sense of passing hours, or seconds. Weariness
     or monotony is economically reinforced.

In using music to be deliberately evocative of a particular time and
place, the producer must be sure of his research. Songs of the First World
War, or ballads of Elizabethan England – there is sure to be at least one
expert listening ready to point out errors of instrumentation, words or
date. To use a piano to set the mood of a time when there was only the
harpsichord and virginals is to invite criticism.

The drama producer must not only search the shelves of the music
library but should sometimes consider the use of specially written material.
This need not be unduly ambitious or costly – a simple recurring folk
song or theme played on a guitar or harmonica can be highly effective.
There are considerable advantages in designing the musical style to suit
the play, and having the music durations to fit the various introductions
and voice-overs.

Production technique

Producers will devise their own methods and different plays may demand
an individual approach, as will working with children or amateurs as
opposed to professional actors. However, the following is a practical
outline of general procedure:

1   The producer works with the writer, or on the script alone rewriting
     for the medium, and making alterations to suit the transmission time
     available.
2   He casts the play, issues contracts, distributes copies of the script,
     arranges rehearsal or recording times.
3   He, or his team, assembles the sound effects, books the studio, arranges
     for any special technical facilities or acoustic requirements, and chooses
     the music.
4   The cast meets, not necessarily in the studio, for a read-through.
     Awkward wordings may be altered to suit individual actors. The
     producer gives points of direction on the overall structure and shape
     of the play and the range of emotion required. This is to give everyone
     a general impression of the piece. Scripts are marked with additional
     information such as the use of cue lights.
5   In the studio, scenes are rehearsed on-mic, with detailed production

points concerning inflection, pauses, pace, movement, etc. The producer should be careful not to cause resentment by 'over-direction', particularly of professional actors. The producer may by all means say what effect he thinks a particular line should achieve. He can, however, undermine an actor if he goes as far as telling him precisely how it should be delivered. Sound effects, pre-recorded or live in the studio, are added. The producer's main task is to encourage the actors and to listen carefully for any additional help they may need or for any blemish that should be eradicated.

6   As each scene is polished to its required perfection, a recording 'take' is made. Are the pictures conjured up at the original reading of the script being brought to life? Is the atmosphere, content and technical quality exactly right? Necessary retakes are made and the script marked accordingly.

7   The recording is edited using the best 'takes', removing fluffs and confirming the final duration.

8   The programme is placed in the transmission system and the remaining paperwork completed.

There are many variations on this pattern of working. Here are alternative approaches:

- Do away with the studio. If the play is suitable, then for example the recording can be made out of doors among the back alleys and railway tracks of the city – a *radio verité*?
- There is no need to think only in terms of the conventional play. A highly effective yet simple format is to use a narrator as the main storyteller with only the important action dialogue spoken by other voices. Few but vivid effects complement this radio equivalent of the strip cartoon – excellent as a children's serial.
- Dialogue on its own can be a simple and powerful means of explaining a point – two farmers discussing a new technique for soil improvement – or illustrating a relationship, such as a father and daughter arguing about what she should wear. Given the right people it doesn't even have to be scripted – give them the basic idea and let them ad-lib it.
- The use of monologue. It sounds dull perhaps, but writers – such as Alan Bennett with his 'Talking Heads' – enable a single character to unravel a richly riveting stream of internal thinking, motivation, musings, and real or imagined behaviour which, spoken in the idiom of the listener, creates immediate identification. Perhaps with a little music and effects added, radio monologue creates an intense world of two – the character and the listener. One such example was the award-winning 'Spoonface Steinberg' – Lee Hall's hour-long internal monologue by a 7-year-old autistic girl, dying of cancer. Just her voice, thinking of the music she loves and why. Imaginative, inspiring, insightful and memorable – a lovely use of the medium.

A producer finding his way into the drama field is well advised to listen to as many radio plays and serials as possible. He or she will gather ideas and recognise the value of good words simply spoken.

In realising the printed page into an aural impression, he must not expect that the visual images in his mind at the start will be exactly translated into the end result. Actors are not puppets to be manipulated at will, they too are creative people and will want to make their own individual contribution. The finished play is an amalgam of many skills and talents, it is a 'hand-made', 'one-off' product which hopefully represents a richer experience than was envisaged by any one person at the outset.

20

Documentary and feature programmes

The terms are often used as if they were interchangeable and there is some confusion as to their precise meaning. But here are exciting and creative areas of radio and because of the huge range which they cover, it is important that the listener knows exactly what is being offered. The basic distinctions of type are to do with the initial selection and treatment of the source material. A documentary programme is wholly fact, based on documentary evidence – written records, attributable sources, contemporary interviews and the like. Its purpose is essentially to inform, to present a story or situation with a total regard for honest, balanced reporting. The feature programme, on the other hand, need not be wholly true in the factual sense, it may include folk song, poetry or fictional drama to help illustrate its theme. The feature is a very free form where the emphasis is often on portraying rather more indefinable human qualities, atmosphere or mood.

The distinctions are not always clear cut and a contribution to the confusion of terms is the existence of hybrids – the feature documentary, the semi-documentary, the drama documentary and so on.

It is often both necessary and desirable to produce programmes which are not simply factual, but are 'based on fact'. There will certainly be times when through lack of sufficient documentary evidence, a scene in a true story will have to be invented – no actual transcript exists of the conversations that took place during Columbus's voyage to the New World. Yet through his diaries and other contemporary records, enough is known to piece together an acceptable account which is valid in terms of reportage. While some compromise between what is established fact and what is reasonable surmise is understandable in dealing with the long perspective of history, it is important that there is no blurring of the edges in portraying contemporary issues. Fact and fiction are dangerous in combination and their boundaries must be clear to the listener. A programme dealing with a murder trial, for example, must keep to the record; to add fictional scenes is to confuse, perhaps to mislead. Nevertheless it is a perfectly admissible programme idea to interweave serious fact, even a court case, with contrasting fictional material, let us say songs and nursery rhymes; but it must then be called a feature not a documentary. Ultimately what is important is not the subject or its treatment, but that we all understand what is meant by the terms used. It is essential that the listener knows the purpose of the broadcaster's programme – essentially the difference between what is true and what is not. If the producer sets out to provide a balanced, rounded, truthful account of something or someone – that is a documentary. If he does not feel so bound to the whole truth and his original intent is to give greater reign to the imagination, even though the source material is real – that is a feature.

The documentary

Very often subjects for programmes present themselves as ideas which suddenly become obvious. They are frequently to do with contemporary issues such as race relations, urban development, pollution and the environment, medical research. A programme might explore in detail a single aspect of one of these subjects which broadly attempts to examine how society copes with change. Other types of documentary deal with a single person, activity or event – the discovery of radium, the building of the Concorde aeroplane, the life of a notable figure or the work of a particular factory, theatre group or school.

Essentially these are all to do with people, and while statistical and historical fact is important, the crucial element is the human one – to underline motivation and help the listener understand why certain decisions were made, and what makes people 'tick'.

The main advantage of the documentary approach over that of the straightforward talk is that the subject is made more interesting and brought alive by involving more people, more voices and a greater range of treatment. It should entertain while it informs, and as it illuminates provoke further thought and concern.

Planning

Following on the initial idea is the question of how long the programme should be. It may be that the brief is to produce for a 30-minute or one-hour slot, in which case the problem is one of selection, of finding the right amount of material. Given a subject that is too large for the time available, a producer has the choice either of dealing with the whole area fairly superficially or reducing the topic range and taking a particular aspect in greater depth. It is, for example, the difference between a 20-minute programme for schools on the life of Chopin, and the same duration or more devoted to the events leading up to Chopin's writing of the 'Revolutionary Study' directed to a serious music audience.

Where no overall duration is specified, simply an intent to cover a given subject, the discipline is to contain the material within a stated aim without letting it become diffuse, spreading into other areas. For this reason it is an excellent practice for the producer to write a programme brief in answer to the questions 'What am I trying to achieve?' 'What do I want to leave with the listener?' Later on, when deciding whether or not a particular item should be included, a decision is easier in the light of the producer's own statement of intent. This is not to say that programmes cannot change their shape as the production proceeds, but a positive aim helps to prevent this happening without the producer's knowledge and consent.

At this stage the producer is probably working on his own, gradually coming to terms with the subject, exploring it at first hand. During this initial research he makes notes, in particular listing those topics within the main subject which must be included. This is followed by decisions on technique – how each topic is to be dealt with. From this emerges the running order in embryo. Very often the title comes much later – perhaps from a significant remark made within the programme. There is no formally recognised way of organising this programme planning; each producer has his own method. By committing thoughts to paper and seeing their relationship one to another – where the emphasis should be and what is redundant – the producer is more likely to finish up with a tightly constructed, balanced programme. Here is an example of the first planning notes for a local radio programme. This radio station serves a coastal region where the trawler fleet has been seriously affected by the loss of fishing rights in international waters:

Working title:	'The Return of the Trawlermen'.
Aim:	To provide the listener with an understanding of the impact which changes in the deep-sea fishing industry over the last ten years have had on the people who work in it.
Duration:	30 minutes.
Information:	Annual figures for shipping tonnage, men employed, fish landed, turnover and profit, investment, etc.
Content:	Historical account of development in the 10 years:
	– technological change, searching, catching, freezing methods.
	– economic change, larger but fewer ships – implication for owners in the increase in capital cost.
	– social change resulting from fewer jobs, higher paid work, longer voyages, and better conditions on board.
	– political change, new regulations relating to international waters, size of nets, off-shore limits.
Key questions:	What has happened to the men and the ships that

	used to work here in large numbers?
	What has happened to those areas of the city where previously whole streets were dependent on fishing as a livelihood? Family life, local shopkeepers, etc., loss of comradeship?
	How significant were the political factors affecting fishing rights in distant waters or was the industry in any case undergoing more fundamental change? Will the trends continue into the future?
Interview sources:	Trawler Federation.
	Docks and Harbour Board? Shipbuilder?
	Trawlermen's Union.
	Fleet owner.
	Skippers – past and present.
	White Fish Authority.
	Representative of fish processing industry – oils – frozen food.
	Government – ministry official; members of parliament.
	Wives of seamen, etc.
Reference sources:	Newspaper cuttings.
	Library – shipping section.
	Government White Paper.
	Trawler company reports.
	Magazine – *Fishing News International*.
	Fred Jones (another producer who did a programme on the docks some time ago).
Actuality:	Bringing in the catch, nets gear running, ship's bridge at sea, engine noise, radio communications, shoal radar, etc.
	Unloading – dockside noises.
	Auction.

By setting out the various factors which have to be included in the programme, it is possible to assess more easily the weight and duration which should be given to each, and whether there are enough ideas to sustain the listener's interest. It probably becomes apparent that there is a lot to get in. It would be possible to do a programme which concentrated solely on the matter of international fishing rights, but in this case the brief was broader and the temptation to dwell on the latest or most contentious issue, such as safety at sea, must be resisted – that's another programme.

There is one final point on planning. A producer's statement of intent should remain fixed, but how he fulfils that aim may change. He plans initially to reach his goals in a certain way; however, if in the course of the production he unearths an unforeseen but vital fact, he must alter his plans to include it. The programme material itself will influence decisions on content.

Research

Having written the basic planning notes, the producer must then make the programme within his allocated resources. He must decide whether he will call on a specialist writer or will write his own script. Depending on this will rest the matter of further research – perhaps it is possible to obtain the services of a research assistant or reference library. The producer who is working to a well-defined brief knows what he wants and in asking the right questions will save both time and money. The principle with documentary work is always as far as possible to go back to sources, the people involved, eye witnesses, the original documents, and so on.

Structure

The main structural decision is whether or not to use a narrator. A linking, explanatory narrative is obviously useful in driving the programme forward in a logical, informative way. This can provide most of the statistical fact and the context of the views expressed, and also the names of various speakers. A narrator can help a programme to cover a lot of ground in a short space of time but this is part of the danger; and may give the overall impression of being too efficient, too 'clipped' or 'cold'. The narrator should *link* and *not interrupt*, and there will almost certainly not be any need to use a narrative voice between every contribution. There are styles of documentary programme which make no use at all of links but each item flows naturally from one to the next, pointing forward in an intelligible juxtaposition. This is not easy to do but can often be more atmospheric.

Collecting the material

Much of the material will be gathered in the form of location interviews. If it has been decided that there will be no narrator, it is important to ensure that the interviewees introduce themselves – 'speaking as a trawler owner . . .' or 'I've been in this business now for thirty years . . .' They may also have to be asked to bring out certain statistical information. This may be deleted in the editing but it is wise to have it in the source material if there is no obvious way of adding it in a linking script.

It must be decided whether the interviewer's voice is to remain as part of the interviews. It may be feasible for all the interviewing to be done by one person, who is also possibly the producer, and for the programme to be presented in the form of a personal investigative report. Pursuing this line further, it is possible for the producer to hire a well-known personality to make a programme as a personal statement – still a documentary, but seen from a particular viewpoint that is known and understood. Where the same interviewer is used throughout, he becomes the narrator and no other linking voice is needed. Where a straightforward narrator is used, the interviewer's questions are removed and the replies made to serve as statements, the linking script being careful to preserve them in their original context. What can sound untidy and confusing is where in addition to a narrator, the occasional interviewer's voice appears to put a particular question. A programme should be consistent to its own structure. But form and style are infinitely variable and it is important to explore new ways of making programmes – clarity is the key.

Impression and truth

The purpose of using actuality sounds is to help create the appropriate atmosphere. More than this, for those listeners who are familiar with the subject, recognition of authentic backgrounds and specific noises increases the programme's authority. It may be possible to add atmosphere by using material from sound effects discs. These should be used with great care since a sound only has to be identified as 'not the genuine article' for the programme's whole credibility to suffer. The professional broadcaster knows that many simulated sounds or specially recorded effects create a more accurate impression than the real thing. The producer concerned not simply with truth but with credibility may use non-authentic sounds only if they give an authentic impression.

The same principle applies to the rather more difficult question of fabrication. To what extent may the producer create a 'happening' for the purpose of the programme? Of course he has to 'stage manage' some of the action. If he wants the sound of ship's sirens, the buzzing of a swarm of angry bees, or children in a classroom reciting poetry, these things may have to be made to happen while his recorder is running. Insofar as these sounds are typical of the actual sounds, they are real. But to fabricate the noise of an actual event, for example a violent demonstration with stones thrown, glass breaking, perhaps even shots being fired; this could too easily mislead the listener unless it is clearly referred to as a simulation. Following the work of broadcasters in wartime it is probably true that unless there are clear indications to the contrary, the listener has a right to expect that what he hears in a documentary programme is genuine material to be taken at face value. It is not the documentary producer's job to deceive, or to confuse, for the sake of effect.

Even the reconstruction of a conversation that actually happened, using the same individuals, can give a false impression of the original event. Like the 'rehearsed interview', it simply does not feel right. Similarly it is possible to alter a completely real conversation by the switching on of a recorder–a house builder giving a quotation for a prospective purchaser is unlikely to be totally natural with a 'live' microphone present!

Faced with the possibility that reality will elude him, both in an original recording or by a later reconstruction, the documentary producer may be tempted to employ secretive methods to obtain his material.

An example would be to use a concealed recorder to get a conversation with an 'underground' bookdealer for a programme on pornography. This is a difficult area which brings the broadcaster into conflict with the quite reasonable right of every individual to know when he is making a statement for broadcasting. Certainly the BBC is opposed to the use of surreptitious production techniques as being an undue invasion of personal liberty. If such a method is used, it is as a result of a decision taken at a senior level.

The implications for an organisation which broadcasts material which depended on the subliminal or secret are such that this is a question which the producer, staff or freelance, should not take upon himself. He must obtain clearance from his programme boss.

Of course if the subject is historical, it is an understood convention that scenes are reconstructed and actors used. Practice in other countries differs but in Britain a documentary on even a recent criminal trial must of necessity reconstruct the court proceedings from the transcripts since the event itself cannot be recorded. No explanation is necessary other than a qualification of the authenticity of the dialogue and action. What is crucial is that the listener's understanding of what he hears is not influenced by an undisclosed motive on the part of the broadcaster.

Music

The current practice is to make little use of music in documentary programmes, perhaps through a concern that it can too easily generate an atmosphere, which should more properly be created by real-life voices and situations. However, producers will quickly recognise those subjects which lend themselves to special treatment. Not simply programmes which deal with orchestras or pop groups, but where specific music can enhance the accuracy of the impression – as background to youth club material, or to accompany reminiscence of the depressed 1930s. A line from a popular song will sometimes provide a suitably perceptive comment, and appropriate music can certainly assist the creation of the correct historical perspective.

Compilation

Having planned, researched, and structured the programme, written the basic script and collected his material the producer must assemble it so as to meet his original brief within the time allotted. First, a good opening. Two suggestions which could apply to the earlier example of the programme on the fishing industry are illustrated by the following script of page one:

Example one

1 Sound effects:	Rattle of anchor chain.
	Splash as anchor enters water.
2 Narrator:	The motor vessel *Polar Star* drops anchor for the last time. A deep-sea trawler for the last twenty-four years she now faces an uncertain future.
	Outclassed by a new generation of freezer ships and unable to adapt to the vastly different conditions, she and scores of vessels like her are now tied up – awaiting either conversion or the scrap yard.
	In this programme we look at the causes of change in the industry and talk to some of the men who make their living from the sea. Or who, like their ships, feel that they too have come to the end of their working life, etc.

Example two

1 Skipper Matthews:	I've been a trawler skipper for eighteen years – been at sea in one way or another since I was a lad. Never thought I'd see this. Rows of vessels tied up like this, just rusting away – nothing to do. We used to be so busy here. I never thought I'd see it.
2 Narrator:	The skipper of the Grimsby *Polar Star*. Why is it that in the last few years the fishing fleet has been so drastically reduced? How have men like skipper Matthews adapted to the new lives forced on them? And what does the future look like for those who are left? In this programme we try to find some of the answers . . . etc.

The start of the programme can gain attention by a strong piece of sound actuality, or by a controversial or personal statement carefully selected from material that is to be heard within the programme. It opens 'cold' without music or formal introduction preceded only by a time check and station identification. An opening narration can outline a situation in broad factual terms, or it can ask questions to which the listener will want the answers. The object is to create interest, even suspense, and involve the listener in the programme at the earliest possible time.

The remainder of the material may consist of interviews, narrator's links, actuality, vox pop, discussion and music. Additional voices may be used to read official documents, newspaper cuttings or personal letters. It is better if possible to arrive at a fairly homogeneous use of a particular technique, not to have all the interviews together, and to break up a long voice piece or statement for use in separate parts. The most easily understood progression is often the chronological one, but it may be desirable to stop at a particular point in order to counter-balance one view with its opposite. And during all this time the final script is being written around the material as it comes in – cutting a wordy interview to make the point more economically in the narration, leaving just enough unsaid to give the actuality material the maximum impact, dropping an idea altogether in favour of a better one. Always keeping one eye on the original brief.

Programme sequence

There are few rules when it comes to deciding the programme sequence. What matters is that the end result makes sense – not simply to the producer, who is thoroughly immersed in the subject and knows every nuance of what he left out as well as what he put in, but to the listener who is hearing it all for the first time. The most consistent fault with documentaries is not with their content but in their structure. Examples of such problems are insufficient 'signposting', the reuse of a voice heard sometime earlier without repeating the identification, or a change in the convention regarding the narrator or interviewer. For the producer who is close to his material it is easy to overlook a simple matter which may present a severe obstacle to the listener. The programme maker must always be able to stand back and take an objectively detached view of his work as its shape emerges.

The ending

To end, there are limitless alternatives. Here are some suggestions:

1   To allow the narrator to sum up – useful in some types of schools programme or where the material is so complex or the argument so interwoven that some form of clarifying resumé is desirable.
2   To repeat some of the key statements using the voices of the people who made them.
3   To repeat a single phrase which appears to encapsulate the situation.
4   To speculate on the future with further questions.
5   To end with the same voice and actuality sounds as those used at the opening.

6   To do nothing, leaving it to the listener to form his own assessment
     of the subject. This is often a wise course to adopt if moral judgements
     are involved.

Contributors

The producer has a responsibility to those whom he has asked to take
part. It is first to tell them as much as possible of what the programme is
about. He provides them with the overall context in which their contribution
is to be used. Second, he tells them, prior to transmission, if their contribution
has had to be severely edited, or omitted altogether. Third, whenever
possible he lets his contributors know in advance the day and time of
transmission. These are simple courtesies and the reason for them is
obvious enough. Whether they receive a fee or not, contributors to
documentary programmes generally take the process extremely seriously,
often researching additional material to make sure their facts are right.
They frequently put their professional or personal reputation at risk in
expressing a view or making a prediction. The producer must keep faith
with them in keeping them up to date as to how they will appear in the
final result.
   What the producer cannot do is to make the programme conditional
upon their satisfaction with the end product. He cannot allow them access
to the edited material in order to have it approved for transmission. Not
only would he seldom have a programme because contributors would not
agree, but he would be denying his editorial responsibility. The programme
goes out under his name and that of the broadcasting organisation. That,
the listener understands, is where praise and blame attaches and editorial
responsibility is not to be passed off or avoided through undisclosed
pressures or agreements with anybody else.

Programmes in real time

A great use of the medium is to unravel the telling of a true story across
the same timespan as the original. It might be a hospital emergency, a
court case, a tropical storm, or a rescue operation. An excellent example
is Len Deighton's 'Bomber', a wartime anniversary broadcast by BBC
Radio 4. It was about a bomber raid over Germany, but it was done in real
time, in chunks over an afternoon and evening. It took us from the initial
lunchtime briefing, the take-off and some of the flight to the raid itself –
by which time everything had gone horribly wrong – and it was interwoven
with the personal details and voices of the people in the air and on the
ground, on both sides, who had done this for real fifty years before. It
was immensely complex, documentary and drama, effects and actuality.
You had to stay up until midnight to hear the planes – some of them –

returning home. Anguish, death, hope, despair, it was innovative as well as authentic – and tremendously compelling radio.

The feature

Whereas the documentary must distinguish carefully between fact and fiction and have a structure which separates fact from opinion, the feature programme does not have the same formal constraints. Here all possible radio forms meet, poetry, music, voices, sounds – the weird and the wonderful. They combine in an attempt to inform, to move, to entertain or to inspire the listener. The ingredients may be interview or vox pop, drama or discussion and the sum total can be fact or fantasy. A former Head of BBC Features Department, Laurence Gilliam, described the feature programme as 'a combination of the authenticity of the talk with the dramatic force of a play, but unlike the play, whose business is to create dramatic illusion for its own sake, the business of the feature is to convince the listener of the truth of what it is saying, even though it is saying it in dramatic form'.

It is in this very free and highly creative form that some of the most memorable radio has been made. The possible subject material ranges more widely than the documentary since it embraces even the abstract. A programme on the development of language, a celebration of St Valentine's day, the characters of Dickens. Even when all the source material is authentic and factually correct, the strength of the feature lies more in its impact on the imagination than in its intellectual truth. Intercut interviews with people who served in the Colonial Service in India mixed with the appropriate sounds can paint a vivid picture of life as it was under the British Raj – not the whole truth, not a carefully rounded and balanced documentary report, it is too wide and complicated a matter to do that in so short a time, but a version of the truth, an impression. The same is true of a programme dealing with a modern hospital, the countryside in summer, the life of Byron, or the romance of the early days of aviation. The feature deals not so much with issues but with events, and at its centre is the ancient art of telling a story.

The production techniques and sequence are the same as for a documentary – statement of intent, planning, research, script, collection of material, assembly, final editing. In a documentary the emphasis is on the collection of the factual material. Here, the work centres on the writing of the script – a strong story line, clear visual images, the unfolding of a sequence of events with the skill of the dramatist, the handling of known facts but still with a feeling of suspense. Some of the best programmes have come from the producer/writer who can hear the end result begin to come together even as he does his research. It is only through his immersion in the subject that he is qualified to present it to the rest of us. Once again, because of the multiplicity of treatment possible and the indistinct definitions we use to describe them, an explanatory subtitle is often desirable.

'A personal account of . . .'
'An examination of . . .'
'The story of . . .'
'Some aspects of . . .'
'A composition for radio on . . .'

Thus the purpose of the finished work is less likely to be misconstrued.
For the final word on the documentary and feature area of programming,
Laurence Gilliam again:

> 'It can take the enquiring mind, the alert ear, the selective eye, and the
> broadcasting microphone into every corner of the contemporary world,
> or into the deepest recess of experience. Its task, and its destiny is to
> mirror the true inwardness of its subject, to explore the boundaries of
> radio and television, and to perfect techniques for the use of the creative
> artist in broadcasting.'

21

The work of the producer

So what does the producer actually do?

Ideas

First and foremost he or she has ideas – ideas for programmes, or items, people to interview, pieces of music or subjects for discussion – new ways of treating old ideas, or creating a fresh approach to the use of radio. New ideas are not simply for the sake of being different, they stimulate interest and fresh thought, so long as they are relevant. But ideas are not the product of routine, they need fresh inputs to the mind. The producer therefore must not stay simply within the confines of his world of broadcasting, but must involve himself physically and mentally in the community he is attempting to serve. It is all too easy for media people to stay in their ivory tower and to form an elite not quite in touch with the world of the listener. Such an attitude is one of a broadcasting service in decline. Ideas for programmes must be rooted firmly in the needs and language of the audience they serve; the producer's job is to assess, reflect and to anticipate those needs through a close contact with his potential listeners. If he is far from them he must read their newspapers, talk to the returning traveller, study his mail most carefully, and visit their country as and when he can. The producer will carry a small notebook to jot down the fleeting thought or snatch of conversation overheard. And if he cannot think of new ideas himself, he must act as a catalyst for others, stimulating, and being receptive to their thoughts and at least recognise an idea when he sees one. Only then may he retreat to the quiet of an office so that he can think.

There is, however, a great deal of difference between a new idea and a good idea and any programme suggestion has to be thought through on a number of criteria. An idea needs distilling in order to arrive at a workable form. It has to have a clarity of aim so that all those involved know what they are trying to achieve. It has to be seen as relevant to its target audience, and it must be practicable in terms of resources. Is there the talent available to support the idea? Is it going to be too expensive in people's time? Does it need additional equipment? What will it cost? Is there sufficient time to plan it properly? Any new programme idea has to be thought through in relation to the four basic resources – people, money, technical equipment and time. It may be depressing to have to modify a really good idea in order to make it work with the resources available, but one of the producer's most important tasks is to reconcile the desirable with the possible.

The audience

Given the initial programme idea in a practical format, the producer may have to persuade his boss, the head of department, programme controller or schedule 'gatekeeper' that the proposal is the best thing that could happen to the broadcast output. Further, not only will the programme not fail, but it will enhance the manager's reputation, as well as provide a memorable programme. While the producer will see a project as intrinsically worthwhile or personally creative, the manager of the service may be much more concerned with competitive ratings. His or her first question is likely to be: 'What will it do for the audience?' There are two possible answers: 'satisfy it' or 'increase it'. A good programme may do both. In allocating a transmission slot, the time of day selected and the material preceding it can be crucial to the success of the programme. It is no good putting out a programme for children at a time when children are not available, nor is it helpful to broadcast an in-depth programme at a time when the home environment is busy and the necessary level of concentration is unlikely to be sustained. This is where a knowledge of the target audience is essential. Farmers, industrial workers, housewives, teenagers, doctors will all have preferred listening times which will vary according to local circumstances.

The fairly superficial news/information and 'current affairs plus music' type of continuous programme, where all the items are kept short, may be suitable for the general audience at times when other things are happening – such as meal times or at work. But the timing of the more demanding documentary, drama or discussion programme can be critical and will depend on individual circumstances. Factors to be considered when assessing audience availability may include weekday/weekend work and leisure patterns, the potential car listenership which can represent a significant 'captive' audience, television viewing habits, FM/MW usage, and so on. The producer is involved in marketing his product and normal consumer principles apply whether or not his radio service is commercially financed. We look at this further in the next chapter.

Resource planning

Having agreed on a time slot for the broadcast, the producer must ensure there is reasonable time available for its preparation. Is it to be next week or in six months' time? No producer will say that he has sufficient time for production work but there is much to commend his working within definite deadlines for such pressure can lend creative impetus to the programme.

The programme idea is now accepted and the transmission date and time allocated. At this stage the producer draws up a detailed budget and obtains authorisation for any additional resources which he may need – money for research effort, scriptwriting, contributors or a music group. Fees may have to be negotiated. He will check the availability of appropriate studio facilities and make arrangements for any necessary engineering or other staff support. He must also obtain the clearance of any copyright work which he wishes to use. Conditions for the broadcasting of material in which usage rights are owned by someone outside the broadcasting service vary widely. In the case of literary copyright, books, poems, articles, etc., the publisher will normally be the point of reference but if the work is not published, the original author (or if dead his estate) should be consulted. Under British copyright law such rights of ownership exist for a period of 70 years from the date of publication or from the death of the author, whichever is the later. The rules vary according to the law of the country in which the broadcast is to be made and in cases of doubt it is well worth taking specialist legal advice – discussing copyright fees after the broadcast is, to say the least, a weak negotiating position.

Preparation of material

Programme requirements may be very simple and the producer able to fulfil them on his own – some interview material, music selected from the library, and his own crisply written and well-presented links may be all that is required. Good ideas are often simple in their translation into radio and can be easily ruined by 'over-production'. On the other hand, it may be necessary to involve a lot more people, such as a writer, 'voices', actors, specialist interviewer or commentator. The interpretation of the original idea may call for specially written music or the compilation of sound effects, electronic or actuality. Again, it is easy to get carried away by an enthusiasm for technique, which is why the original brief is such an important part of the process. It should serve as a reference point throughout the production stages.

After selecting his contributors, agreeing fees and persuading them to share his objectives, the producer's task is basically to stay in close touch with them. Remember also that the producer must always be on the look-out for new voices and fresh talent. He then revises draft scripts and clarifies individual aims and concepts so that when everybody comes together in the studio they all know what they are doing and can work together to a common goal. He must apply himself to this with a great sense of timing so that everything integrates at the same moment – the broadcast or recording. Above all the producer must give encouragement. The making of programmes has to be both creative and businesslike. There is a product to be made, restrictions on resources and constraints of time to be observed. But it also calls on people to behave uniquely, to write something they have never written before, to give a new public performance, to play music in a personal way. The producer is asking them to give something of themselves. Contributors, artists and performers of all kinds generally give their best in an atmosphere of encouragement, not uncritical, not complacent, but with a recognition that they are involved in the process of creative giving. To an extent it is self-revealing, and this leaves the artist with a feeling of vulnerability which needs to be reassured by a sense of succeeding in his attempted communication. The producer's role is to provide this feedback in whatever form it is required. He has therefore to be watchful and perceptive of his contributors, whether they are professional or amateur.

During this time while material is being gathered and ordered there may be a number of permissions to seek. Broadcasters have no rights over and above those of any other citizen and to interview someone or to make recordings in a home, hospital, school, factory or other non-public place requires the approval of appropriate individuals. In the great majority of cases it is not withheld, and indeed it is most often only an informal verbal clearance that is required. It does not do, however, to record on-site without the knowledge or consent of the legitimate owner or custodian of the property. But neither is it acceptable to be given permission subject to certain conditions, for example to undertake to play back the material recorded and not to broadcast any of it without the further permission of the person concerned. In response to his request to record, the producer must accept only a 'yes' or a 'no' and not be tempted to accept conditional answers. The listening public has the right to believe that the programme they hear is what the producer whose name attaches to it wants them to hear, and is not the result of some secret deal imposed on him by an outside party. Accountability for the programme rests with the producer: it can seldom be shifted elsewhere.

So the programme takes shape – promising ideas developed, poor material discarded, items and thoughts explained, put into the listener's context, juxtaposed to give variety, impact, chronology or other meaningful structure. In its design the producer must remember to engage the listener's attention at the start of the programme and continue to do so throughout. Now comes the time for it to be broadcast or recorded.

The studio session

Here again the producer must combine his talents for shrewd business with his yearning for artistic creativity. He has limited resources, particularly of time, and he has people wanting to give of their best, some of whom may be in unfamiliar surroundings, possibly tense, almost certainly nervous. So, arriving in good time he must set them at their ease and create the appropriate atmosphere. There is no single 'right' way of doing this since people and programmes are all different – the atmosphere in a news studio needs to be different from that of a drama production. A music recording session will be different again from a talk or group discussion. It may be a case of providing coffee all round or even a *small* quantity of something stronger. Any lavish hospitality of this kind is generally much better left until after the programme. But there are two points which a producer must observe. First he must make any necessary introductions so that people know who everyone else is and what they are doing, including any technical staff. The second task is to run over the proposed sequence of events so that individual contributors know their own place in the timescale. These two practices help to reassure and provide some security for the anxious. There is nothing worse for a contributor than his standing around wondering what is going on or even whether he has come to the right place.

The producer has brought everything needed for the programme; pre-recorded inserts correctly labelled and timed with optional out cues; music inserts timed; plenty of copies of the script; coloured pens for making changes; stopwatch, and so on. Everyone in the studio should have a script or running order, and know what is required of them. They should know of any breaks in the session and be sufficiently acquainted with the building as to be able to find their own way to the source of coffee or to the lavatory. There should be enough chairs. Rehearsal, recording or transmission can begin.

Whatever the attitude and approach of the producer, it will find its way into the end product. To get the job done, there has to be a certain studio discipline. 'That's not quite right yet, let's take it again from the beginning' – this is the signal for a new and better concerted effort. 'Everyone check the running order' – means *everyone*. 'Start again in twenty minutes' cannot mean people wandering back in half an hour. The producer needs to control, to drive the process forward, to maintain the highest possible quality with the time and talent at his disposal. It is generally a compromise. Too strict a control can be stifling to individual creativity, anxiety increases, the studio atmosphere becomes formal and inflexible. On the other hand, lack of control can mean a drifting timescale, an uncertainty as to what is going on, and a lowering of morale. The appropriate balance is developed with experience, but the following points apply generally when managing a studio full of people:

1   Use general talkback for announcements to studio participants sparingly. Such use should be brief and should be overall praise or straightforward administration. Never use talkback from the control cubicle into the studio for individual criticism.

2   Listen to suggestions from contributors for alterations but be positive
    in making up your mind as to what will be done.
3   Provide plenty of individual feedback to contributors.
4   Keep in the mind the needs of the technical, operational or other
    broadcasting staff – they also want to feel that they are contributing
    their skills to the programme.
5   Watch the clock, plan ahead for breaks, recording or transmission
    deadlines. Avoid a last-minute rush.
6   Mark the script as a recording proceeds for any retakes needed or
    editing required.
7   If rehearsing for a live programme, work out and write down any
    critical timings for particular items. The 'must be finished by' times
    are most critical.
8   If the programme is live and is underrunning or overrunning – and
    you can do little about it, for example a concert – tell the person who
    needs to know as far in advance as possible and agree what is to
    happen.
9   Be encouraging. Be communicative. Keep calm. Keep control.

   In rehearsing a straight talk, it may be necessary for the producer to sit
in the studio opposite the speaker in order to persuade him that he is
actually talking to someone. The effect of knowing that he has an intent
listener is likely to make his delivery much more natural. Moreover, any
verbal pedantry or obscure construction in the script is the signal for the
producer to ask for clarification. Since it is given in conversational form,
this can then become the basis of the suggested rewrite. Almost always,
constructive suggestions for simplification, professionally given, are
gratefully accepted, often with relief. Producers should remember, however,
that their role is not to create in their contributors imitations of themselves.
In making suggestions for script changes, or how an actor might tackle a
certain line, the producer must be visualising not how he himself would
do it but how that particular performer can be most effective.
   In the presence of a live mic, or through a glass window, the producer's
non-verbal language is characterised by the following most universal
hand signals:

1   The cue for someone to start. The hand is brought from the raised
    position to point directly at the person to speak. This is also used for
    handovers from one broadcaster to another.
2   To keep an item going, e.g. to lengthen an interview. The hands are
    slowly moved apart as if stretching something between them.
3   To start winding up. The index finger describes slow vertical circles
    in the air, getting faster as the need to stop becomes more urgent.
4   To come to an immediate end, to cut. The hand is passed swiftly
    across the throat – often with an anguished facial expression.

So much for the producer's responsibility to the other people involved in making the programme but of course his prime responsibility is towards the listener. Is the programme providing a clear picture of what it is intending to portray? Are the facts correct and in the right order? Is it legally all right? Is it of good technical quality? Is it interesting? Most of these questions are self-evident and so long as they are borne in mind will answer themselves as the programme proceeds. Some questions, however, may require a good deal of searching, for example – is it in good taste?

Taste

Taste: The disposition or execution of a work of art, choice of language or conduct seen in the light of the faculty of discerning and enjoying beauty or other excellence.

(Concise Oxford Dictionary)

Beauty and excellence are in the eye of the beholder but where there are common standards of acceptability, the producer must be fully aware of them so that his programmes may be directed within them. He must know the generally accepted social tenor of his time and the cultural flavour of his place if he is to succeed with the general audience and avoid giving unwitting offence. He may decide that his programme is only directed to the bawdy revellers in the marketplace with little thought for those who would be shocked at such goings on. He may design his programme simply for a cultural or intellectual elite whose acceptable standards are 'more advanced' than those of more simple folk. So be it, but either way the radio casting of his programme is by definition broad rather than narrow; others will hear and their reaction too must be calculated as part of the overall response. In questions of content, material will be designed for a specific target audience but the matter of acceptable taste is a much broader issue which the radio producer must sense accurately. In stepping outside it, he takes a considerable social risk. In deciding on a style of language, or the inclusion of a particular joke which raises the question of good or bad taste, there is one simple rule: would I say this to someone I did not know very well in a face-to-face situation? If so, it is fine for broadcasting. If not, the producer must ask himself whether he is using the microphone as a mask to hide behind. Simply because the studio appears to be isolated from contact with the audience it is sometimes tempting to be daring in one's assumed relationship with the individual listener. The seeming separation is not a cause for bravado but a reason for sensitivity. The matter of taste in broadcasting so often resolves itself in a recognition of the true nature of the medium.

If in real doubt over a knife-edge decision, it is wise to talk to a senior more experienced colleague. For this reason the matter of taste is discussed further in Chapter 22, The executive producer.

Ending the session

After a recording and while the contributors are still present, it is often possible to put together some additional material for on-air trailing and promotional use. A specially constructed 30-second piece will later pay dividends in the attention which it can attract.

The producer has a responsibility to his professional colleagues who use the same technical facilities. This finds expression in a number of ways:

1 *Studio cleanliness:* the smaller the radio station, the more it operates on a 'leave the place as you would wish to find it' basis. He probably will not be required to do the clearing up in detail but he should leave it in its 'technically normal' and usable state.
2 *Fault reporting system:* every studio user must contribute to the engineering maintenance by reporting any equipment faults which occur. It is extremely annoying for a producer to be seriouly hampered by a studio 'bug' only to find that someone else had the same trouble a few days previously but did nothing about it.
3 *Return of borrowed equipment:* a radio centre is a communal activity, its facilities are shared. An additional disc player or special microphone taken from one place to another for a specific programme should be returned afterwards. It may not be the producer who actually does this but it is likely to be his responsibility to ensure that some other user is not deprived.

If the programme was 'live', the contributors have been thanked and the occasion suitably rounded off. This may mean the dispensing of some 'hospitality' or simply a discussion of 'how it went'. It is generally unhelpful to be too analytical at this stage: most people know anyway whether as a programme it was any good.

Post-production

If the programme was recorded, the producer may have to do some editing. He should have a running order or script marked in detail with the edits he knows he wants but there may be additional cuts to be made in the light of the overall timing. A further editing session is booked at which he listens to all the material and makes the final judgement on what is to be included. There may still be time here for second thoughts. Should the music be re-mixed? Does the sound need to be enhanced in any way – by adding echo or special effects treatment? This is probably the last opportunity to hear the programme in its final form to check that what the listener hears is what the producer intends.

Programme administration

The finished programme together with the necessary paperwork is then deposited within 'the approved system' so that it finds its way satisfactorily on to the air. Often, a producer while excellent as a creative impresario, artistic director or catalyst in the community may have a total blindspot when it comes to simple programme administration. He may be fortunate enough to have a secretary to look after much of this for him, nevertheless it is his responsibility to see that such things are done. The following is a summary of the likely tasks:

1   The completion of a recording or editing report and other details such as library numbers which will enable the programme to get on the air in accordance with the system laid down.
2   The writing of introductory on-air announcements, cues and other presentation material detailing the transmission context of the programme.
3   The initiation of payment to contributors, and letters of thanks giving the transmission details if these were not known at the time of the recording.
4   The supply of programme details covering the use of music or other copyright material. Depending on local circumstances these items will need to be reported to the various copyright societies so that the original performers and copyright holders can receive their proper payment.
5   The issue of a publicity handout, press release, or programme billing for use by newspapers, or programme journal published by the broadcasting organisation. The placing of on-air trails or promos drawing attention to his programme.
6   The answering of correspondence generated by the broadcast. While not necessarily representative of listener reaction as a whole, letters form an important part of a producer's public accountability. Apart from the PR value to the particular radio service such enquiries and expressions of praise or criticism constitute a consumer view which should not be treated lightly.

Technician, editor, administrator and manager

In summary, the producer's task is in four parts: the technical and operational, the editorial, the administrative, and the managerial. The technical part is to do with the proper use of the tools of his trade, knowing when and how to use programme-making equipment. The editorial function is about ideas and decisions. It is to do with making judgements about what is and is not appropriate for a particular programme. It is about backing hunches and taking risks, about choosing and commissioning material. The administrative part is procedural (following agreed systems of paperwork): contracts, running orders and scripts, expenses and payments, overtime, leave applications, studio bookings, copyright returns, logging transmissions, verifying traffic, reporting faults, requisitioning music and Fx. But the producer is also a manager, managing projects called programmes. He sets the objectives for other people, monitors their progress, controlling, organising and motivating them in their work. He will be the person who disciplines the habitual latecomer, who resolves conflict between contributors.

The journalist and the DJ, the presenter and the performer, frequently regard themselves as the pre-eminent component in a mixed sequence. The producer as manager must create the team where each is sufficiently confident to support the other. As manager, the producer recognises the financial responsibility of the job – agreeing the budget, monitoring expenditure, and taking action to remain within the allocation. If need be he argues the case for more, but the editorial and managerial aspects cannot be separated. Editorial decisions *are* resource decisions. Like any other manager he is primarily in charge of the quality of what happens, his bottom line is the standard of the programme – he says what is good enough and what isn't. At the end of the day the producer decides and communicates what he wants done, to what standard, by when, by whom, at what cost. That's editorial management.

Having completed the programme, the producer is already working on the next. For some it is a constant daily round to report new facts and discover fresh interests. For others it may be a painstaking progress from one epic to another. Unlike the purely creative artist, the producer cannot remain isolated, generating material simply from within himself. His role is that of the communicator, the interpreter who attempts to bring about a form of contact which explains the world a little more. For the most part it is an ephemeral contact leaving an unsubstantial trace. Radio works very much in the present tense, reputations are difficult to build and even harder to sustain. The producer is rarely regarded as any better than his or her last programme.

22

The executive producer

Senior producer, programme manager, controller, organiser, or director, the producer's boss comes in a wide range of guises and titles. His or her essential job is not so much to make programmes as to make programme makers. Whether responsible for a programme strand, a station, or a network, this senior editorial figure is there to listen to the output and provide feedback on it to the producers. A handwritten note left on the producer's desk – a habit of former BBC Managing Director of Radio, David Hatch – is particularly effective. This head of programmes will also lead the programme meetings where there is a systematic discussion of the output – there is more detail on this procedure in the next chapter. The purpose of this is continual improvement through professional encouragement and evaluation, critique, discussion of ideas, programme development, and so on. Without this, the output stagnates.

This senior role also has the job of 'gatekeeper' – the person who says yes or no to programme suggestions, who takes risks with new ideas, who spots talent and encourages new presenters – or gets rid of them. In his or her mind all the time are questions such as:

- What is the purpose of this station?
- What are my success criteria?
- Are the current programmes achieving these?
- If not, how can I improve – the programmes?
                              – the presenters?
- What stops the output being better in my terms?
- How can I overcome this?
- What is my competition doing?

The executive producer needs to find answers by combining the two aspects of the job – the people and the task. To concentrate solely on the people might make for good relationships but what is the quality of the work? On the other hand, to put the whole emphasis on task may create an unduly work-driven atmosphere in which no one really thrives. The balance is seen in a management style which bears fruit in the creative work done by motivated people who enjoy being part of a team.

Scheduling

Some would describe the drawing up of a good programme schedule as perhaps the highest art form in radio. Certainly there are schedulers who seem to have just the right knack of placing programmes and people where the target audience most appreciates them.

The basis of a schedule that works is (1) deciding the role and purpose of the station – *what do I want to say, to whom, and with what effect?* and (2) knowing the needs, likes and dislikes, habits, work patterns, and availability of the intended audience.

The first of these comes from perceiving a gap in the market and supplying what is needed – rock music, pop, classical, news, speech-based programming, phone-in chat, community services, etc. No station can do everything and deciding the limits of programme provision is crucial. The second is discovered by audience research in whatever form possible. There may be an organisation set up to do this and it is a question of buying the information needed. In the absence of such a facility, it is necessary either to commission research or to infer the information from other sources, e.g. government statistics, newspaper circulation figures, population breakdown by socio-economic groups, etc. The scheduler ideally knows what the target audience is doing at different times of the day – and night – and at the weekend. What time the intended audience wakes up and goes to bed. When the home of the intended audience is busy – typically at breakfast time – so as to keep items short, with plenty of time checks. When the home is quiet and the listener available for more extended listening. When the typical listener is likely to be travelling – by car – 'drivetime'. When the time is right for news or relaxation. How to capitalise on a popular programme by following it with something really appropriate for the inherited audience.

Audience habits and therefore listening habits are very different at the weekends. Broadcasters to Muslim communities know that this mostly applies to Friday and Saturday, while Sunday is generally a normal working day. The schedule also needs to recognise special days – anniversaries, holidays, the New Year, Christmas and Easter, Hanukkah, Ramadan and other religious festivals.

Television schedules make much use of library material – feature films or repeats. Radio may also be able to do this depending on the format. For example, a live programme recorded off-air in the evening may be usefully repeated for a different audience in the morning, or during the night, in the same week. Schedulers have to remember that habitual listeners to any programme do not like change – a change of presenter for a holiday break is understandable, but a new presenter may take months to be accepted. A change of timing for a programme may mean that the regular listener cannot then hear it at all. This should therefore only be done after extensive and conclusive research. Continual tinkering with the schedule simply irritates, giving rise to the cry from the audience – 'who do they think this station belongs to? – a good question.

Rescheduling

Sooner or later an event occurs which will cause the cancellation of the planned schedule. This may happen in one of three ways.

First, in the short term, following the death of a national leader or other eminent person it may be appropriate to play 'solemn music'. This is especially so when the death is unexpected or is violent. But what does such music mean within the context of the station's format? How should the station react to the assassination of a head of state as opposed to the anticipated natural death of an ageing president? Standby music of an appropriate kind should always be available for such emergencies. However, with the best will in the world it may turn out to be quite wrong – like the music station which on the death of the Pope found that its first standby track was a non-vocal version of 'Arriverderci Roma'!

The second, longer-term, rescheduling applies either when a death requires a longer period of mourning or in the event of a major national disaster such as a serious terrorist incident or outbreak of war. This becomes not simply a matter of solemn music, but of revising the whole output, perhaps to allow for additional news slots, and of changing programmes. This was certainly the case in the days following the death of Diana, Princess of Wales, and in India after the assassination of Prime Minister Gandhi. It is also necessary to advise presenters on how to be sensitive to the public mood, for it is here that off-the-cuff remarks or anarchic humour can cause offence among the audience.

The third circumstance for rescheduling requires special vigilance. It most frequently occurs after a one-off event of significant national or local impact – a tragedy at sea or an air crash, a particularly brutal killing or terrorist bomb. In itself the event may not justify the cancellation of the planned schedule – but these things have a habit of coinciding with something in the output which is quite inappropriate in the new context. Programme titles – even song titles – have to be scrutinised to avoid the unfortunate juxtaposition.

Strategic planning

One of the roles of senior editorial staff is to give a sense of movement to the output – that this year is not the same as last, but things have moved on. The station is not only up to date in the sense of being topical, but it is also advancing in its technology, its skills, and its relationship with the listener.

Strategic planning requires a response to such questions as:

Where do I want this station to be in 5 years' time?
What do I want to be known for?
What should we be doing that's new?

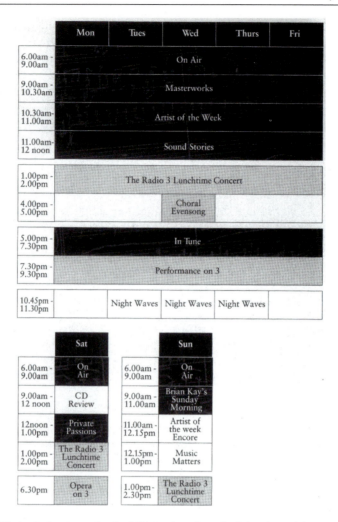

	Mon	Tues	Wed	Thurs	Fri
6.00am - 9.00am	On Air				
9.00am - 10.30am	Masterworks				
10.30am- 11.00am	Artist of the Week				
11.00am- 12 noon	Sound Stories				
1.00pm - 2.00pm	The Radio 3 Lunchtime Concert				
4.00pm - 5.00pm			Choral Evensong		
5.00pm - 7.30pm	In Tune				
7.30pm - 9.30pm	Performance on 3				
10.45pm - 11.30pm		Night Waves	Night Waves	Night Waves	

	Sat
6.00am - 9.00am	On Air
9.00am - 12 noon	CD Review
12noon - 1.00pm	Private Passions
1.00pm - 2.00pm	The Radio 3 Lunchtime Concert
6.30pm	Opera on 3

	Sun
6.00am - 9.00am	On Air
9.00am - 11.00am	Brian Kay's Sunday Morning
11.00am - 12.15pm	Artist of the week Encore
12.15pm - 1.00pm	Music Matters
1.00pm - 2.30pm	The Radio 3 Lunchtime Concert

Figure 22.1 A simple method of illustrating the schedule – for publicity purposes – when the daily programmes are stripped across the week, with variations at the weekend. (Courtesy BBC Radio 3)

It is armed with a sense of vision that the strategic planner begins to glimpse new programme ideas and in discussion with others slowly turns them into reality. So, we have a 3-5 year plan not simply in the business sense but one that envisions programme development.

Even with a single programme strand it is possible to think strategically. For example, an educational series is an obvious case of leading the listener to a point at the end that was not the same as the beginning. The idea is to cover a body of knowledge or skill that the listener acquires – understanding a period of history, or learning a language. But this concept of 'taking the listener somewhere' need not be confined to formal education, it applies to all leadership, i.e. of having a vision of where you want to get to, and enabling the listener to get there. Competitive quiz programmes begin with initial rounds, and semi-finals, culminating in a final. DJ music programmes can be designed so that they move towards a live concert or festival.

In such ways the output, although consistent, is not the same week after week, and year after year. It is changing and developing, and extending the audience as it does so – not in a haphazard, accidental way but as part of a planned strategy. Even the public servant may exercise such leadership.

Commissioning programmes

A radio organisation may not produce all its programmes in-house. It may approach an outside production facility or programme supplier to fill part of its schedule. How then does it ensure that it gets the programming it wants – that the programme will be appropriate for the audience listening to the station at that time?

The commissioning of programmes varies widely in different parts of the world – not least in the matter of money, of who pays whom. Some organisations will specify what is wanted and pay a freelance producer or independent production facility to supply it. Others will offer air-time to a programme maker who then buys it in order to reach the audience which the broadcaster is targeting – the producer recouping the cost from a sponsor or other donor agency. The danger here is that the schedule may become a patchwork of unrelated, independent programmes with little or no continuity. However, even when air-time is sold, it is possible to create a coherent schedule through a positive commissioning procedure.

Retaining editorial control means that the executive producer has to specify with some precision what is wanted. Programmes are commissioned within the framework of a known and agreed editorial policy. The purpose of the programme or series has to be stated, not in great detail, but sufficiently to evaluate the result in terms of its intended audience and desired response.

For example, a farming programme in East Africa might be specified as follows:

A 30-minute programme in the Hausa language designed for the working community to enable them to make more efficient use of the land, and to promote self-sufficiency through improved crops by the better use of simple tools, seeds, and other agricultural products, in line with government policy. The speech-based programme format may include scripted talk, interview, discussion, or drama. No commercial sponsor shall dictate programme content.

The specification for a comedy programme in the UK could be:

The purpose of this late Saturday evening programme is to amuse and

XYZ RADIO   PROGRAMME COMMISSIONING FORM

Date: _____

Programme to be supplied by:

Address:

Programme title: _____   No. of progs _____

Service/block: _____   Language _____

Transmission: Day _____   Time _____   Duration_____

Purpose of the programme:
_____

Primary target audience:
_____

Programme format(s):
_____

Specification of content:
_____

Limits of commercial sponsorship:
_____

Cost/budget:
_____

Goals for listener response:

Procedure for follow-up, correspondence, etc.

Technical specification:

Programme delivery:

Copyright considerations:

Other requirements:

This programme commission will be reviewed on _____

Signed for XYZ RADIO _____

Signed for Programme Supplier _____

entertain especially through the satirising of the events of the previous week in the world of politics and publicly reported news. The format may include talk, sketches, songs, dialogue, etc. using a variety of voices. The programme must be self-regulating in matters of libel, good taste, etc.

Essential details regarding cost, technical specification, and programme delivery must also be negotiated. To ensure that the agreement is as watertight as possible, a 'Programme Commissioning Form' is drawn up. This will be given to the programme supplier together with any other statements of editorial policy, and a legal contract.

The BBC, when commissioning programmes for Radio 4, is very keen that programme suppliers describe what the listener will hear – not simply the proposed programme content, but how it would sound on radio. To help with correctly identifying the target audience, each part of the day has a detailed commissioning brief, with a profile of the audience at that time – the age, proportion of men and women, social groups, and what they are likely to be doing – together with editorial guidelines as to the kind of programme wanted. The Programme Commissioning Form illustrated on pp. 284–5 is a simple basis for the procedure used in an African context.

Complaints

When the broadcaster is thought to be 'wrong' the listener tends to complain in one of three categories:

1   On hearing something which affects the listener personally, such as what they believe to be an incorrect news report in which they were involved, or a lack of balance in reporting a particular argument. They are aggrieved at being treated unfairly.
2   A historical inaccuracy, a wrong date, or a misquote – an error of fact.
3   When they are offended by something said – bad language, overtly sexual comment, inappropriate humour – a matter of taste.

While all media attract the inveterate letter writer, especially seeming eccentrics, a station which takes its audience seriously should regard what a listener says with some importance. People like a feeling of having some control – at least that their complaints are taken note of, that their views matter. They therefore need a reply, either by letter or on the air in the same time slot as the item which caused their offence. An error of fact requires an apology and a correction, a difference of view requires an explanation while at the same time respecting the other opinion.

In the UK, the Broadcasting Standards Commission has its own Codes of Guidance – on Fairness and Privacy, on Taste and Decency and general programme standards. While mostly to do with television, these are useful documents of which all broadcasters should be aware. They raise issues such as the need to avoid suffering and distress when interviewing people involved in personal tragedy, the use of hidden microphones, about not causing offence against religious sensibilities – or the bereaved. They outline the care needed in areas such as drugs, alcohol, race, gender stereotypes, people with disabilities or mental health problems, etc. They draw attention to the lyrics of songs which glamorise crime or invite aggression.

There is always a balance to be struck between the broadcasters' right to freedoms of speech and expression and the listeners' right not to have their own liberties infringed. While programmes should be creative and perhaps experimental, challenging convention, they also need to recognise their responsibility to different audiences. Radio is not for graffiti on-air. It is not an anonymous communication, but always signed and attributed.

Research indicates that listeners and viewers have clear images in their minds about the channels and stations they use. Many complaints arise because a programme was at odds with their expectation of it – a family programme was found to be offensive or shocking, or a provocative programme did not sufficiently challenge or provoke. The complaint is often not in the deed but in the unexpected or uncharacteristic. A station, a programme, must live up to its promise. It must satisfy – if possible exceed – the expectations the listener has of it. In the event of serious complaint, the executive role is to calm the anger or allay the fears of the listener; it is to defend the producer, and deal internally with the cause.

Codes of Practice

The executive producer is the guardian of the programme output, ensuring that it does not fall foul of the legal necessities placed on broadcasters. But it is not simply libel and race relations and the other laws of the land which have to be considered, there is often a mass of regulatory constraints or Codes of Practice – some voluntary, some not – which come with a station's licence to broadcast.

In the UK, the Radio Authority – the licensing body for the independent radio sector – produces a Programme Code which overlaps with the BSC Code already referred to, but which is written specifically for radio. Distinguishing between rules and advice, the Code usefully lists the subject areas which need special vigilance. The Broadcasting Act (1990) says

> 'nothing shall be included in programmes which offends against good taste or decency or is likely to encourage or incite to crime or to lead to disorder or to be offensive to public feeling'.

The problem of course is in the interpretation of this requirement – it is difficult to challenge or provoke without offending someone. No one wants blandly soporific radio. The Code attempts to define and pinpoint likely matters of contention:

- Sexual explicitness
- Blasphemy
- The portrayal of violence
- Bad language especially in programmes for younger listeners
- The danger of misleading people through simulated news bulletins and reconstructions
- The interviewing of children or criminals
- Appeals for donations
- The whole area of the occult
- Rules for competitions, and so on.

The Authority also publishes Codes on 'News and Current Affairs', and on 'Advertising and Sponsorship' (see p. 122). The BBC has its own 'Producer Guidelines' – in the form of an interactive CD and a book. Suffice it to say that since ignorance of the law is no defence, senior programmers should acquaint themselves with this part of the professional infrastructure – designed as much for the protection of broadcasters as for the public. Knowing where the boundaries are means that it is possible, when necessary, to test and challenge them.

Involving the audience

Producers in their programmes have several ways of involving the audience and we have looked at some of these earlier (see pp. 191–2) – phone-ins, competitions, letters, etc. There are plenty more.

- Sponsorship of a travelling drama/music group
- Running on-air courses with a correspondence element
- Establish a Birthday Club for children
- Set up a network of 'focus groups' to discuss and report back on programmes
- Invite programme contributions
- Raise money by asking for donations to a specific cause
- Set up a website
- Establish access for e-mail
- Ask for listeners' jokes or personal stories
- Offer a free telephone number to call for help with specific topics – finance, AIDS, unemployment, marriage/parenting problems
- Run an unfinished drama/story for the listener to complete.

Beyond this, there is a broader scale of involvement which is part of the strategic outreach of the station – national or local – which falls within the senior production role. These events may be geared to audience research, to find out about likes and dislikes, or needs, or it may be for publicity value simply to promote the station. In either case the aim is to extend the listenership in size, reach, or quality. These events include:

1 The travelling roadshow – a popular programme takes to the road, broadcasting each time from a different venue, generally 'live' in front of an audience.
2 The radio conference – station management, producers, and presenters host a question-and-answer session with the public, distribute and collect questionnaires on listening habits, ask for and discuss ideas for future output, and sell merchandise. These events are probably not broadcast although the questions and answers may be recorded for transmission to demonstrate public service involvement.
3 The one-off major event, music festival or rock concert – sponsored by the station, perhaps with non-radio co-sponsors. This may be a commercial enterprise at an established venue, charging for admission and broadcasting the performance.
4 The national scriptwriting competition – for story, radio drama, poetry. Using eminent judges, it may be sponsored with considerable spin-off publicity and press coverage.
5 The Radio Exhibition – illustrating its own history, to mark a special event, involving other appropriate organisations, sponsors, equipment manufacturers, providing an insight into the future. Held at its own premises, at a public venue, or in a prestigious retail store, staff and presenters are on hand to give out publicity, answer questions, sell merchandise, and with a studio facility, produce inserts into station output.

Events such as these involve a station well beyond its normal sphere of influence and resources. They require a great deal of planning, and frequently considerable up-front capital costs. This is why they need senior management intervention. Nevertheless for a station that wants to make its mark, to celebrate its anniversary, or to contribute to public life in a memorable way, the executive producer must look beyond the day-to-day schedule to the long term, asking how best he may raise the profile of the station among his potential listenership.

Website

Senior station staff have to decide what kind of presence they want on the Internet. Undoubtedly, the dissemination of the audio output in this way is a part of broadcasting as is distribution by cable – in fact there are stations which use the Internet as their only means of transmission.

The complexity of a site is governed not so much by the amount of information provided or the number of linked pages but by the rate at which page content is changed. Categories of increasing complexity and therefore input effort and cost are as follows:

1   A single static page with details of station name, logo, frequency, address, coverage area, and purpose statement.
2   A page that is updated periodically, e.g. to show variations in the weekly schedule. It may also carry advertising.
3   Page(s) that are changed continually to give the latest news and weather together with other information, using text and symbols.
4   A website that provides an audio output of the most recent news bulletin and news updates, plus station promos and announcements.
5   Provision of the continuous live audio output.
6   A fully interactive site which offers keyboard text discussion with users in real time.

The possibilities for website design are virtually limitless. They include links to presenter pages, sound clips of recent news output – with pictures, a live camera image of the studio, a tour of the station, special announcements, coming events, classified ads – jobs and cars, chart music listings, details of music played, interactive opinion-seeking questions, advice pages, etc.

The more complex the design, especially with regard to pictures and animated graphics, the longer it will take to download, and therefore the greater the potential for frustration and the user not completing the connection. When this occurs, a surfer is unlikely to return to the site to use it. There is therefore much to be said for making the first entry page relatively straightforward giving the basic information and a menu of links, designed in bold colours with succinct text reversed out on a dark background.

Inputting data – text, graphics and audio – is a relatively cheap process, but the greater the rate at which it is changed, the faster and more expensive must be the data connection with the service provider. Before starting a website, the webmaster should explore other radio websites, be prepared to negotiate over costs and if necessary seek specialist advice.

Archival policy

• What of the output should be kept?
• In what form?
• How should it be catalogued?

Decisions regarding the archives inevitably fall to the senior programme staff – producers themselves have a tendency to want to keep everything.

It may be that the whole of the station/network output is automatically kept on a slow-speed logging recorder. However, the quality of these recordings, although good enough for technical questions of content, are seldom up to rebroadcasting. In any case they generally have a limited life before they are erased and rerecorded.

But why keep archives anyway? This will depend considerably on the nature of the output. A music station may want to keep very little, a news station potentially at least will want to keep most things in the short term since any story has the capacity to grow. Keeping news items for a week or two is sensible, either to follow a developing story or to produce a pull-together of the week's news. If the news is held on computer hard disk a capacity of 150–200 hours (mono) is generally sufficient. However, to avoid the disk running out of space, old material must be cleared and items for keeping downloaded to some other medium such as the floopy disk, VHS cassette, or recordable CD. These must then be clearly marked with the start and end dates of the material recorded. The data will carry the accompanying text and cue material but the physical marking or cataloguing of the recording must clearly indicate what it contains, or at least the period covered. It is best to keep an archive database capable of search by date, subject, title, and producer/presenter name. It should not be forgotten that the purpose of an archive is not storage but retrieval.

Over the years many archives have built up a substantial number of programmes recorded on $\frac{1}{4}$-inch tape. If it is impractical to convert these to a digital medium, special care is needed in their long-term storage. The enemies of any magnetically sensitive medium, including audio, DAT and VHS cassettes, are excessive heat, vibration or banging, and external electrical or magnetic fields, for example from some microphones. Each of these affect the magnetic elements on the tape causing loss of quality, particularly of the upper frequencies – in extreme cases partial or total erasure. For the long-term storage of tape the recommended temperature is 10°C (50°F).

A characteristic of tape in storage is its remarkable anonymity. A CD or disc is inseparable from its label but one piece of tape is precisely like another. A spool, cassette, or cartridge – and its box – should carry sufficient information for the programme material to be readily identified. Ideally this will include:

- Subject matter
- Name and address of speaker, musicians, performers, etc.
- Location of recording
- Date of recording
- Details of copyright material
- Duration
- Catalogue number or category.

Items designed for the archives should be labelled in this way before archiving as a matter of routine, whatever the recording medium.

Good decisions about archives and a sensible system will mean that repeats, year-end programmes, anniversaries, and other retrospectives are easy to produce and satisfyingly representative. The station then deals not only with the ephemera of today but becomes the repository of the community or nation – the guardian of its oral history.

23

Programme evaluation

A crucial activity for any producer is the regular evaluation of what he or she is doing. Programmes have to be justified. Other people may want the resources, or the airtime. Station owners, advertisers, sponsors and accountants will want to know what programmes cost and whether they are worth it. Above all, conscientious producers striving for better and more effective communication will want to know how to improve their work and achieve greater results. Broadcasters talk a great deal about quality and excellence, and rightly so, but these are more than abstract concepts, they are founded on the down-to-earth practicalities of continuous evaluation.

Programmes can be assessed from several viewpoints. We shall concentrate on three:

- Production and quality evaluation.
- Audience evaluation.
- Cost evaluation.

Production evaluation

Programme evaluation carried out among professionals is the first of the evaluative methods and should be applied automatically to all parts of the output. However, it is more than simply a discussion of individual opinion, for a programme should always be evaluated against previously agreed criteria.

First, the basic essential of the proper technical and operational standard. This means there is no audible distortion, that intelligibility is total, that the sound quality, balance and levels are correct, the fades properly done, the pauses just right and the edits unnoticeable.

Second, is about what the programme is for – what does it set out to do? A statement of purpose should be formulated for every programme so that it has a specific direction and aim. Without such an aim, any programme can be held to be successful. So, what is the 'target audience'?

What is the programme intending to do for that audience? How well does it set about doing it? (Whether it actually succeeds in this is an issue for audience evaluation.)

Third, is a professional evaluation of content and format? Were the interviews up to standard? The items in the best order? The script lucid, and the presenter communicative? These questions are best discussed at a regular meeting of producers run by the senior editorial figure. It is important that everyone has heard the programme under review – either off-air or played at the start of the meeting. It is good to have some positive feedback first – what did we like about it? In what ways did we feel it appealed to the target audience? What could be improved? Is there a follow-up? Everyone should be encouraged to take part in this evaluation, including of course the programme's producer. The focus of the discussion should be on constructive improvement rather than finding fault. When producers are first involved in playback and discussion sessions, they are bound to show some initial defensiveness and sensitivity over their work. This has to be understood. It is best minimised by focusing the discussion on the programme, not on the programme maker. The process is essentially about problem solving and creatively seeking new ideas in pursuit of the programme aim.

Programme quality

Quality is a much over-used word in programme making. Is it only something about which people say 'I know it when I see it or hear it, but I wouldn't like to say what it is'? If so, it must be difficult to justify the judges' decisions at an awards ceremony. Of course there will be a subjective element – a programme will appeal to an individual when it causes a personal resonance because of experience, preference or expectation. But there must also be some agreed professional criteria for the evaluation of programme excellence. In a programme of quality at least some of the following will be prominently in evidence.

First, *appropriateness*. Irrespective of the size of the audience gained, did the programme actually meet the needs of those for whom it was intended? Was it a well-crafted piece of communication which was totally appropriate to its target listeners, having regard to their educational, social or cultural background? Programme quality here is not about being lavish or expensive; it is about being in touch with a particular audience, in order exactly to serve it, providing with precision the requirements of the listener.

Second, *creativity*. Did the programme contain those sparks of newness, difference and originality which are genuinely creative, so that it combined the science and logic of communication with the art of delight and surprise? This leaves a more lasting impression, differentiating the memorable from the dull or bland.

Third, *accuracy*. Was it truthful and honest, not only in the facts it presented and in their balance within the programme, but also in the sense of being fair to people with different views? It is in this way that programmes are seen as being authoritative and reliable – necessary of course for news, but essential also for documentary programmes, magazines and, in its own way, for drama.

Fourth, *eminence*. Quality acknowledges known standards of ability in other walks of life. A quality programme is likely to include first-rate performers – actors or musicians. It will make use of the best writers and involve people eminent in their own sphere. This of course extends to senior politicians, industrial leaders, scientists, sportsmen and women – known achievers of all kinds. Their presence gives authority and stature to the programme. It is true that the unknown can also produce marvels of performance, but quality output cannot rely on this and will recognise established talent and professional ability.

Fifth, *holistic*. A programme of quality will certainly communicate intellectually in that it is understandable to the sense of reason, but it should appeal to other senses as well – the pictorial, imaginative or nostalgic. It will arouse emotions at a deeper and richer level, touching us as human beings responsive to feelings of awe, love, compassion, excitement – or even the anger of injustice. A quality programme makes contact with more of the whole person.

Sixth, *technical advance*. An aspect of quality lies in its technical innovation, its daring – either in the production methods or the way in which the audience is involved. Technically ambitious programmes, especially when 'live', still have a special impact for the audience.

Seventh, *personal enhancement*. Was the overall effect of the programme to enrich the experience of the listener, to add to it in some way rather than to leave it untouched – or worse to degrade or diminish it? The end result may have been to give pleasure, to increase knowledge, to provoke or to challenge. An idea of 'desirable quality' should have some effect which gives, or at least lends, a desirable quality to its recipient.

Eighth, *personal rapport*. As the result of a quality experience, or perhaps during it, the listener will feel a sense of rapport – of closeness – with the programme makers. One intuitively appreciates a programme that is perceived as well researched, pays attention to detail, achieves diversity or depth, or has personal impact. The listener identifies not only with the programme and its people, but also with the station. Programmes which take the trouble to reach out to the audience earn a reciprocal benefit of loyalty and sense of ownership.

Combining accuracy with appropriateness, for example, means providing truthful and relevant news in a manner which is totally understandable to the intended audience at the desired time and for the right duration. Quality news will also introduce creative ways of fairly describing difficult issues so leaving the listener feeling enriched in his understanding of the world.

Programme quality requires several talents. It takes time to think through and is less likely to blossom if the primary requirement is quantity rather than excellence. It cannot be demanded in every programme, for creativity requires experiment and development. It needs the freedom to take risks and therefore occasionally to make mistakes. Qualitative aspects of production are not easy to measure, and it may be that this is why an experienced programme maker determines them intuitively rather than by logic alone. Nevertheless they have to be present in any station which has quality on the agenda, or aspires to be a leading broadcaster.

Quality allied to programming as a whole – especially that thought of as public service – takes us back to criteria described on p. 11. Quality in this sense will mean a diversity of output, meeting a whole range of needs within the population served. It will reflect widely differing views and activities, with the intention of creating a greater understanding between different sections of the community. Its aim is to promote tolerance in society by bringing people together – surely always the hallmark of quality communication.

The cynic will say that this is too idealistic and that broadcasting is for self-serving commercial or even propagandist ends – to earn a living, and provide music to ease the strain of life for listeners. If this is the case, then simply evaluate the activity by these criteria. The many motivations for making programmes and the values implicit in the work are outlined on p. 15. What is not in doubt is the need to evaluate the results of what we do against our reason for doing it.

Audience evaluation

Formal audience research is designed to tell the broadcaster specific facts about the audience size and reaction to a particular station or to individual programmes. The measurement of audiences and the discovery of who is listening at what times and to which stations is of great interest not only to programme makers and station managers but also to advertisers or sponsors who buy time on different stations. Audience measurement is the most common form of audience research, largely because of the importance attached to obtaining this information by the commercial world.

Three methods of measurement are used and in each, people are selected at random from a specific category to represent the target population:

1   People are interviewed face to face, generally at home.
2   Interviews are conducted by phone.
3   Respondents complete a listening diary.

The more detailed the information required – the number of people who heard part or all of a given programme on a particular day, and what they thought of it – the more research will cost. This is because the sample will need to be larger, requiring more interviews (or more diaries) and because the research has to be done exclusively for radio. If the information required is fairly basic – how many people tuned to Radio 1 for any programme last month and their overall opinion of the station – the cost will be much less since suitable short questions can be included in a general market survey of a range of other products and services.

It should be said that constructing a properly representative sample of interviewees is a process requiring some precision and care. For example, we know that the unemployed are likely to be heavy users of the media generally, yet as a category they are especially difficult to represent. Nevertheless a correct sample should cover all demographic groups and categories in terms of age, gender, social or occupational status, ethnic culture, language, and lifestyle, e.g. urban and rural. It should reflect in the proper proportions any marked regional or other differences in the area surveyed. This pre-survey work ensures, for example, that the views of Hindi-speaking students, male and female, are sought in the same ratio to the over-65s living in rural areas as these two categories exist in the population as a whole. Only when the questioning is put to a correctly constructed sample of the potential audience will the answers make real sense.

A further important definition relates to the meaning of the word 'listener'. Many sequence programmes are two or three hours long – do we mean a listener is someone who listened to it all? If not, to how much? In Britain, RAJAR – Radio Joint Audience Research – conducts surveys of all radio listening using a self-completion diary. Here a listener is defined as someone who listens for a minimum of 5 minutes within a 15-minute time segment. The figure for weekly reach is the total number of such listeners over a typical week, expressed as a percentage of the population surveyed. The resulting figures have even more value over a period of time as they indicate trends in programme listening, seasonal patterns, and changes in a specific audience, such as car drivers. This allows comparisons to be made between different kinds of format and schedule. Research therefore helps us to make programme decisions, as well as providing the figures for managers to justify the cost of airtime.

Research panels

Another method of research is through listening panels scattered throughout the coverage area. Such groups can, by means of a questionnaire, be asked to provide qualitative feedback on programmes. Panel members will be in touch with their own community and therefore may be chosen to be broadly representative of local opinion. Once a panel has been established, its members can be asked to respond to a range of programme enquiries which over time may usefully indicate changes in listening patterns.

Such panels are also appropriate where the programme is designed for a specific minority such as farmers, the under-fives, the unemployed, hospital patients, adult learners, or a particular ethnic or language group. Here, the panel may meet together to discuss a programme and provide a group response. Visited by programme makers from time to time, a panel can sustain its interest by undertaking responsible research for the station. But beware any such sounding-board which is too much on the producer's side. Too close an affiliation creates a desire to please, whereas programme makers must hear bad news as well as good. Indeed, one of the key survey questions is always to find out why someone *did not* listen to my programme.

Questionnaires

In designing a research questionnaire ensure that the concepts, words and format are appropriate to the person who will be asked to complete it. A trained researcher filling in the form while undertaking an interview can cover greater complications and variables than a form to be completed by an individual listener on their own. Before large-scale use, any draft questionnaire should be tested with a pilot group to reveal ambiguities or misunderstandings. Here are some design criteria:

• Decide exactly what information you need, and how you will use it.
• Do not ask for information you don't need – redundant questions only complicate things.
• Write the introduction to indicate who wants the information, and why, and what will be done with it. Establish the level of confidentiality.
• Number each question for reference.
• The layout should be in lines rather than boxes or columns. This enables it to be completed by typing or longhand.
• The information you may want is in three categories
   – Facts: name, age, family, address, job, newspapers read
   – Experience: listening habits, reception difficulties, use of TV/videos
   – Opinion: views of programmes, presenters, of competitor stations
• Questions here come in four categories
   – Where the answer is either yes or no: Are you able to receive Radio XYZ?   Yes/No
   – Where you offer a multiple choice question:
     How difficult is it to tune your radio to Radio XYZ?   very difficult/moderately difficult/fairly easy/very easy
   – Where you provide a numerical scale for possible answers: On a scale of 0 – 5 how difficult is it to tune your radio to Radio XYZ? Very difficult 0 . . . 1 . . . 2 . . . 3 . . . 4 . . . 5 Very easy
   – Where you invite a reply in prose:
     What difficulties have you had in receiving Radio XYZ?

Radio XYZ

1. Programme title ....

2. Date/time of broadcast ....

3. Do you listen to this station? (circle)
     every day/most days/once a week/once a month/never

4. Did you hear this particular programme? (circle)   yes/no

5. If no, go to Question 11. If yes, did you listen to (circle)
     all of it/most of it/parts of it/a little of it?

6. What did you think of the programme? (circle)
     excellent/good/fair/poor

7. What did you like most about it?

8. What did you dislike about it?

9. Is the programme broadcast at a suitable time for you? (circle)   yes/no

10. If no, what would be a better time for you? (circle)
      on the same day/on a different day

11. If you did not hear the programme, why was this?

12. If you did hear the programme did you do anything as a direct result?

Some information about yourself is very useful since it enables us to contact you for follow-up if necessary, and it helps us to know more about our audience. However, this section is optional and you may leave it blank if you prefer. Any information you give is for the purpose of programme evaluation only and is regarded as strictly confidential.

Name .... Sex M/F

Address .... Married/Single

....

Age (circle) under 15/16–24/25–39/40–59/60+

Occupation ....

Hobbies and interests ....

....

Other Radio Stations listened to ....

Newspapers/magazines read: ....

Figure 23.1 A simple audience research questionnaire for individual members of a listening panel. The station completes 1 and 2 before distribution

- Yes/No, tick box, multiple-choice, and numerical scale questions take up little space and can be given number values and so are useful in producing statistics.
- Add 'other' to appropriate multiple choice questions to allow for responses you have not thought of.
- Prose answers can be difficult to decipher but give good insights and usable quotes.
- Avoid imprecise terms – 'often', 'generally', 'useful', other than in a multiple-choice sequence.
- Avoid questions that appear to have a right or preferred answer.
- Keep the questionnaire simple, and as short as possible.

Needless to say the whole area of audience research is too big to be dealt with fully here. Further sources of information are listed under Further reading.

Letter response

Informal audience research – anecdotal evidence, press comment and immediate feedback – often has an impact on the producer which is out of proportion to its true value. Probably the most misleading of these – in relation to the listenership as a whole – is the letter response. Several studies have shown that there is no direct correlation between the number of letters received and the size or nature of the audience. News is often the most listened-to part of the output, yet the newsroom receives comparatively few letters.

Letters will indicate something about the individuals who are motivated to write – where they live, their interests perhaps, what triggered them to pick up a pen, or what they want in return. But it is wrong to think that each writer represents a thousand others – they may do, but you don't know that and cannot assume it. There may be more letters from women than from men – does that indicate that there are more women listeners than men? Not necessarily – it may be that women have more time, are more literate, are more motivated, or have the stamps!

Broadcasting to 'closed' countries, or where mail is subject to interference, frequently results in a lack of response which by no means necessarily indicates a small audience. Low literacy or a genuine inability to pay the postage are other factors which complicate any real accuracy in attempting audience evaluation through the correspondence received. It may give some useful indicators, and raise questions worthy of feeding back into programmes – for each individual letter has to be taken most seriously – but letter writers are self-selecting along patterns which are likely to have more to do with education, income, available time and personal motivation than with any sense of the audience as a whole.

A method which partly overcomes the unknown and random nature of letter response is to send with every reply to a correspondent a questionnaire (together with a stamped addressed return envelope) designed to ask about the writer's listening habits, to your own and to other services. It also asks for information about the person. Over a period of several months it is possible to build up some useful demographic data. It still only relates to those who write, but it can be compared with official statistical data – available from many public libraries – to discover how representative are the people who write.

Cost evaluation

What does a programme cost? Like many simple questions in broadcasting this one has a myriad possible answers, depending on what you mean.

The simplest answer is to say that a programme has a financial budget of 'X' – an amount to cover the 'above the line' expenses of travel, contributors' fees, copyright, technical facilities, and so on. But then what is its cost in 'people time'? Are staff salaries involved – producer, technical staff, secretarial time? Or office overheads – telephone, postage, etc.? Is the programme cost to include studio time, and is that costed by the hour to include its maintenance and depreciation? And what about transmission costs – power bills, engineering effort, capital depreciation?

Total costing will include all the management costs and all the overheads over which the individual programme maker can have little or no control. One way of looking at this is to take a station's annual expenditure and divide it by the number of hours it produces so arriving at a cost per hour. But since this results in the same figure for all programmes no comparisons can be made. More helpful is to allocate to programmes all cash resource costs which are directly attributable to it, and then add a share of the general overheads – including both management and transmission costs – in order to arrive at a true, or at least a truer, cost per hour figure which will bear comparison with other programmes.

Does news cost more than sport? How expensive is a well-researched documentary, or a piece of drama? How does a general magazine compare with a phone-in or a music programme? How much does a 'live' concert really cost? Of course it is not simply the actual cost of a programme which matters – coverage of a live event may result in several hours of output, which with recorded repeat capability could provide a low cost per hour. Furthermore, given information about the size of the audience it is possible, by dividing the cost per hour by the number of listeners, to arrive at a cost per listener hour. So is this the all-important figure? No, it is an indicator among several by which a programme is evaluated.

Relatively cheap programmes which attract a substantial audience may or may not be what a station wants to produce. It may also want to provide programmes which are more costly to make and designed for a minority audience – programmes for the disabled, for a particular linguistic, religious or cultural group, or for a specific educational purpose. These will have a higher cost per listener hour but will also give a channel its public service credibility. It is important for each programme to be true to its purpose – to achieve results in those areas for which it is designed. It is also important for each programme to contribute to the station's purpose – its Mission Statement.

After a full evaluation, happy is the producer who is able to say 'My programme is professionally made to a high technical standard, it meets precisely the needs of the whole audience for which it is intended, its cost per listener hour is within acceptable limits for this format, and it contributes substantially to the declared purpose of this station.' Happy too the station manager.

24

Training

We learn in four main ways: by watching others, by studying theory, by trying things out, and by full-scale practical experience. In fact, the activities are linked together. The process of observation leads us to draw conclusions about what appears to work and what doesn't, we can then test theories before launching out in practice. We monitor ourselves as we do that and the cycle starts again.

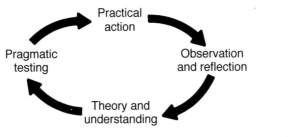

(after Kolb)

Unfortunately, most people are not equally disposed to these four ways of learning. Someone strong on observation, reflection and theory will learn a great deal from visiting studios, watching professionals at work, asking questions and reading the literature – and they may be reluctant to try a practical exercise until they think they have mastered the theory. Someone else – the practical activist – will be anxious to get on with 'doing it', impatient of the principles.

It follows that a good training scheme for producers should contain all four elements in proportions suited to the participants and, in practice, to the facilities and expertise available. There's no point in throwing someone in at the deep end ('that's the only way to learn, that's how I did it') without the essential guidelines. Nor is it sensible to insist that new producers watch someone else without any opportunity to ask questions and try things out for themselves. Giving 'on-air' experience too soon can induce in some a real sense of fear, from which recovery may be slow and painful. On the other hand, preventing people from doing anything 'for real' is likely to lead to acute frustration. It is this aspect of motivation that those responsible for a producer's training must monitor most closely. Is the new person enjoying learning? How do they think the training process is going? Are there signs of their being 'over-challenged' or bored? If they are slow to finish work, is it because they are aiming at too high a standard for their present level of skill? If the work quality is low, is it because of a lack of understanding of what is required or an inability with a particular technique? Learning *how* to interview, for example, will certainly take place during practice and real interviews – but the key to *understanding* interviewing will occur during a session of expert feedback on the end result. The critical analysis of programme material by someone experienced in the craft is an essential part of the learning process. Neither is this confined to the new producer. Established producers also need to grow; their development should not be overlooked simply because they can do a day's work without supervision. New challenges, techniques, programme formats and roles will help to keep regular output producers from going stale. They may even be good enough to enthuse and train the next generation.

Triggers for training

Training used to be regarded as a one-off event experienced early on in a career. However, with continual technological and organisational change, it has to be a lifelong process. For trainers, the events which trigger training include the following:

- New legislation
- The organisation's appraisal process
- High staff turnover
- New equipment or procedures
- A high error rate, or listener complaints
- New programme methods
- Expansion of the broadcasting service
- Multi-skilling of staff and freelances
- New staff joining, or existing people being promoted
- Downsizing, where staff assume greater responsibilities
- New markets, programme areas or services
- Changed organisational or departmental structures.

Trainers will be continuously aware of new training needs, but in a truly innovative organisation senior management will be the driver of visionary change, asking the training department to help implement it.

Learning objectives

What is important in all this of course is not the training but the learning.

What is the intended effect on the learner's attitude, knowledge, or skill – and does it happen? We should therefore start with a clear idea of what we are trying to do by writing a series of statements which describe the intentional outcomes of the course.

For example, at the end of a radio news production course an already trained journalist will be able to:

- Select news stories appropriate to the target audience.
- Put stories in order of importance and write a programme running order.
- Write bulletin material to time using accurate, clear, and engaging language.
- Research news leads, and follow up existing stories.
- Commission and brief reporters to cover news stories, including those remote from the newsroom.
- Package news material, interviews and actuality in audio form in appropriately creative ways.
- Produce and run a studio for a live news transmission.
- Present news on-air.
- Analyse and critique a news programme, giving professional feedback to contributors.

Once the trainer is sure of the skills – or competences – that have to be mastered, the course can be designed by breaking it down into individual sessions, each with its own objectives. Training outcomes, applied even to a single session, describe what the training is designed to achieve in terms of what the student will either know, or be able to do. The success of the training/learning process can then be assessed against these objectives.

Course organisation

The trainer is the enabler, developer and promoter of others. To be successful it is necessary to know the needs of the trainees – their present levels of skills and knowledge – and where you want a particular training event to take them. Whether that event is a half-day module or a three-month course, it will require five areas of attention:

1   *Aim.* The purpose of the training should be clear. What insights and abilities are people to take away with them? The trainer needs a vision of what is to be achieved, and from what starting point.
2   *Logistics.* Technical, financial and other practical arrangements have to be made for the desired number of people attending the event. Teaching space, accommodation, working space together with equipment for practical sessions, administrative effort and office support, e.g. word-processing and copying, transport, catering, visual

aids – OHP, whiteboards, flipcharts, books and folders, etc. – have to be predicted and provided.

3 *Design.* What topics need to be covered in what order? The flow should accommodate the different learning styles, the balance between theory and practical, and between individual and group work. The sessions after a mid-day meal should either succumb to the convention of siesta or be vigorously participative, and anything in the evening should be different again. Courses are frequently so full of input that there is little time for reflection, to process what is learned. Lectures should be modified with discussion 'buzz groups' and questions, and practical work given time for debriefing and individual feedback, encouragement and critique.

4 *Lesson plans.* Each session needs its own written outline starting with its aim. What is it to achieve? How will it do it? A plan, preferably drawn up by the trainer who is to lead the session, will detail the content of the session giving approximate timings for each section and how it will be organised. It will list the training handouts to be distributed, equipment needed, aural and visual aids required, videos, films, cassettes or sound clips to be used, etc.

5 *Evaluation.* The initial aim will give you the success criteria for the training, but has it worked for each individual? Described in detail later, evaluation of and by trainees and tutors is part of the quality control process – and if there are standards to be met and examinations passed, these should also be with regard to the four learning styles.

The following training ideas can be adapted to suit specific conditions and represent the principles of learning by seeing – understanding – trying – doing. Observation, theoretical principle, group discussion, and working in the 'safety' of a training environment are combined with 'on-job' learning. In each case the trainer should clearly indicate what is required, by when, to what standard, with what resources.

Stretching imagination

Write a one-minute piece on a colour (see p. 125). Members of a group take colours out of a hat – black, purple, red, grey, etc. Provide access to music, poetry, Fx and compile a recording for discussion and evaluation by the group. Invite a blind person to sit in and comment.

Editorial selection

Provide each trainee with the same recording of a five-minute interview, to be edited down to $1\frac{1}{2}$ minutes. Use a transcript to mark what has been used and what cut. With a group, each person says why he or she chose certain parts and omitted others, and which parts should be rewritten as cue material. Analyse the reasons for selectivity. Does everyone stand by their own decision, or recognise that other choices may be better?

News priorities

List twelve basic stories. Choose three as the lead stories for a five-minute bulletin. With only room for nine, decide which three to drop. Analyse and discuss the reasons given:

1 Police chief presses for stricter measures against all forms of civil terrorism including up to 28 days' detention without charges being made.
2 Farmers fear price rises of staple foods in the coming six months due to poor harvesting conditions.
3 Popular national youth movement announces plans for international rally to be held in the capital.
4 The country as a whole has attracted more tourists than ever before. Income from tourism reaches multi-million record.
5 Important political figure – in the opposition party – claims wasteful government spending on road-building programme.
6 Plane crash in desolate region involving internal flight of domestic airline with 75 people on board. Circumstances and casualties not yet known.
7 Famous local sportsman wins premier prize in an international competition.
8 Country's political leader announces new government policy for welfare facilities for the old, disabled and poor.
9 Small bomb explosion in a store in centre of the capital. Part of a continuing campaign of protest by a dissident minority who claims responsibility by phone call to the radio station.
10 Rural region suffers a suspected outbreak of cattle disease which threatens the destruction of livestock under a government order.
11 Industrial dispute over a pay claim threatens the shut-down of major car plant.
12 University department of medical research announces a breakthrough in its search for a drug to alleviate arthritis in the hand and knee joints.

News exercise

An excellent 'real' news exercise is to provide a training group with the same sources as those available to the working newsroom. Alongside the professional team the trainees independently compile a five-minute bulletin for comparison with the actual output. Invite the editor to listen and comment – discuss the differences of selection and treatment.

A further exercise is to listen on the same day to bulletins from different stations or networks, analysing the reasons for the variations between them.

Voicework

When giving feedback – especially critical feedback – it is always wise to remember that the trainer is commenting on the work, *not* the person. However, with voice training the work and the person come very close together and it can be almost impossible to separate them. Listening to a trainee newsreader you may feel that he or she is acting, not being themselves. So who are they being? Do they have a mental image of a newsreader that they must live up to? Are they in effect impersonating a newsreader? If so, what is wrong with being him or herself? These can be difficult questions which in the end only the individual concerned can answer. Before going on to specific technical skills any newsreader or presenter must be comfortable with themselves. If they do not like their voice or accent they will try to change it and the whole effect may sound false; for if they are too concerned with their own performance they will not have sufficient care for the listener, and communication will suffer.

The first step therefore is to record and playback some newsreading and ask the reader to comment on it. If it is very different from their normal voice, record this conversation and play this back and compare it with the newsreading. Why the difference? Many readers have to be assured that their normal voice – or at least something very close to it, that works for them throughout the greater part of the day – is perfectly OK for radio.

Having liberated a voice to be natural, rather than either assumed or nervous, the trainer continues with professional feedback according to the following '7 Ps':

1 *Posture*. Is the sitting position comfortable, to allow good breathing and movement? Cramped or slouching posture does not generally make for an easy alertness.
2 *Projection*. Is the amount of vocal energy being used appropriate to the programme?
3 *Pace*. Is the delivery correct? Too high a word rate can impair intelligibility or cause errors.
4 *Pitch*. Is there sufficient rise and fall to make the overall sound interesting? Too monotonous a note can quickly become very tiring to listen to. However, animation in the voice should be used to convey natural meaning rather than achieve variety for its own sake.
5 *Pause*. Are suitable silences used intelligently to separate ideas and allow understanding to take place?

6   *Pronunciation.* Can the reader cope adequately with worldwide names and places? If a presenter is unfamiliar with people in the news, or musical terms in other languages, he must be taught to read phonetic guidelines.
7   *Personality.* The sum total of all that communicates from microphone to loudspeaker, how does the broadcaster come over? Is it appropriate to the programme? Can the trainer or trainee suggest any improvement.

All voicework has something of performance about it and it is natural for broadcasters of all disciplines to want some form of professional feedback. It follows that the opportunity for formal voice training, and for a discussion of one's personal approach to it, should be offered to both new and experienced practitioners.

Personal motivation

Using the lists in Chapter 1, write a short essay on the ideal use of the radio medium and how you see yourself being most fulfilled as a programme maker. Why and how do you want to use radio?

Vox pop

Trainees produce edited street interviews on a similar topic:

- A film or television programme
- An aspect of industrial or agricultural policy
- Solutions to traffic problems
- Views on new buildings being erected
- Teenage dress.

Discuss – to what extent does luck play a part in the end result? How does one's personal approach affect the outcome? What is the most appropriate slot for these items?

Commentary

An initial exercise is for a trainee to be given a picture taken from a newspaper or magazine and without showing it to the others is asked to describe it for 30 seconds. He or she then shows it to the group – they then comment on the differences between their mind's-eye picture and the real one. What was it about perspective, size or content that was distorted? The next step is to record a piece of outside scene-set or event commentary for subsequent playback (without editing) and analysis. Did it provide a coherent picture?

Drama

Write ten minutes of dialogue for two to four voices, with or without effects. Using actors, produce the playlet for discussion by the group. Invite the actors to give their view of how they were produced. Could they have done better? Were the producer's instructions clear? Did the end result re-create the writer's intentions?

New challenges for old producers

Set out deliberately to do something personally never attempted before:

- A vox pop in an old people's home, or at a school.
- Produce a commercial, and offer it to the appropriate agency for comment.
- Draft outline ideas for a documentary, giving research sources.
- Using the clock format, construct a one-hour music sequence for a given target audience.
- List ten new ideas for an afternoon magazine.
- Write a public service message – road safety, community health.
- Reconstruct the station's morning schedule.

Without attempting to justify the result, it should be discussed with an experienced broadcaster. It should then be done a second time, with a further appraisal.

Maintaining output

Many radio training courses with access to studio facilities will culminate with a sequence of programmes as simulated continuous output. Run in real time to a predetermined schedule, a full morning's music, news and weather, traffic reports, features, etc. are watched by professional observers who report back on programme quality, presentation style, sound levels, operational faults, the management of the studio, the producer's ability to motivate and communicate with others, and so on. Some such exercises include deliberate emergencies like equipment failure, the breaking of a major news story, the unexpected arrival of a VIP, or the loss of some pre-recorded material. Such 'disasters' should not be allowed to bring everything to a halt since the effect may be counter-productive to good morale. However, 'live' broadcasters must be encouraged to think quickly 'on their feet'.

Assessing quality

Listen to a piece of radio for subsequent discussion. What was its purpose? What effect did it have on you? Comment on content, order, and presentation – the message and its style of delivery, on technique, story treatment, music selection, news values, etc. If possible invite the actual producer to comment on the discussion. When giving feedback – and this is a general rule – comment on the work, don't criticise the person. Critique the programme not the programme maker.

The process of discussion, analysis and evaluation is carried on continuously by professional broadcasters. It is especially important after any programme that has attracted public or government criticism. Communicators need the comments of others – and a 'gut' reaction may be just as valid as a careful intellectual assessment. All broadcasters, in training and the trained, need to maintain their own analytical reasoning in good order. Keep 'quality' on the agenda.

Training evaluation

If programmes are to be evaluated, then so must be the training process itself. Trainees can be encouraged to set their own goals at the start of any course or training period, and asked half-way through how they are doing in achieving those goals. At the end, to what extent have they succeeded in meeting their own criteria? Other 'end of course' feedback useful to trainers is obtained by questions such as:

- Which sessions did you find most useful?
- Which sessions did you find least useful?
- How did you find the balance of theory and practical? too much theory/ about right/too much practical.
- Did the training come too soon/about the right time/too late, for you?
- How far was the training relevant to the work you are doing/you expect to be doing?
- What would you like to see added to the course?

It is also useful in an end of course questionnaire to ask for 'any other comments'. This will cover training administration and group relationships, as well as course content. The trainer needs to know as much as possible about what the trainee feels about the training experience as well as what he thinks.

Of course radio training in all its forms is only a means to an end –

better broadcasting. Real evaluation can only take place three or six months later, involving both the trainee and his or her manager. Is what has been learned being put to practical use? Are genuine results apparent from the training effort? If not, is there a mismatch between the training approach and the workplace? Training has to meet the felt needs of the programme output; and the trainer, like the programme maker, will constantly evaluate what he does in order to improve the product for the customers.

25

Back-announcement

Harry Vardon, one of the great exponents of golf, was asked why he never wrote a book setting out all that he knew about it. His reason was that when he came to put it down on paper it looked so simple that 'anyone who didn't know that much about it should not be playing the game anyway!'

It looks as though it is the same with broadcasting. Is there really any more to it than – 'Have something to say, and say it as interestingly as you can?' Yet there are whole areas of output which have hardly been mentioned – educational programmes, light entertainment and comedy, programmes for young people, specialist minorities or ethnic groups. What about the special problems of short-wave broadcasting, or programmes for the listener who is a long way from the broadcaster? A book, like a programme, cannot tell the whole truth. What the reader, or listener, has a right to expect is that the product is 'sold' to him in an intelligible way and then remains true to his expectations. Broadcasters talk a great deal about objectivity and balance, but even more important, and more fundamental, is the need to be fair in the relationship with the listener. A broadcasting philosophy which describes itself in programme attitudes yet ignores the listener is essentially incomplete.

What then is the purpose of being in broadcasting – the aim of it all? It is not enough to say, 'I want to communicate' – communicate what? And why? As we have seen there are several possible answers to this – to earn money, to meet the needs of the organisation, to meet your own needs, to become famous, or to persuade others to think as you do. But the purpose of communication is surely to provide options and a consequent freedom of action for other people, not to close them by offering a half-truth or by weighting them with a personal, political or commercial bias. The reason for providing information, education and the relaxation of entertainment is to suggest alternative courses of action, to explain the implications of one against another, and having done so to allow for a freedom of thought and action. This assumes that people are capable of responding in a way which itself requires a regard for our fellow men and women.

Having announced his intentions and made his programme, the producer must put his name to it. Programme credits are not there simply to feed the ego or as a reward for your labours. They are a vital element in the power which broadcasting confers on the communicator – personal responsibility for what is said. Many members of a team may contribute to a programme, but only one person can finally decide on the content. Good programmes cannot be made by committee. Group decisions inevitably contain compromise and a weakening of purpose and structure, but worse, they conceal responsibility. Communication that is not labelled or attributed is of little use to the person who receives it. And if your programme is savaged by a critic or scorned by detractors, first check it by the standards of quality referred to earlier, then remember the following 'anonymous' comment.

'It is not the critic who counts; not the man who points out where the strong man stumbled or where the doer of deeds could have done better. The credit belongs to the man who is actually in the arena; whose face is marred by dust and sweat and blood; who strives valiantly, who errs and comes short again and again; who knows the great enthusiasms; who, at best, knows the triumph of high achievement; and who, at the worst, if he fails, at least fails while daring greatly, so that his place shall never be with those cold and timid souls, who know neither victory nor defeat.'

Programme makers face a hundred difficulties not mentioned here but by engaging a 'professional overdrive', rather than regarding them as a personal undermining, most problems can be made to take on more the aspect of a challenge than a threat. The practicalities of production are encapsulated by the well-known Greenwich time signal whose six pips must serve as their final reminder:

1 *Preparation:* state the aims, plan to meet them.
2 *Punctuality:* be better than punctual, be early.
3 *Presentation:* keep the listener in mind.
4 *Personal:* come over as yourself, be real.
5 *Punctilious:* observance of all agreed systems and procedures. If there is something you don't like, do not ignore it, change it.
6 *Professional:* putting the interests of the listener and the broadcasting organisation before your own. And the constant maintenance of an editorial judgement based on a full awareness and a competent technique.

Glossary

Above-the-line cost   Expenditure under the producer's control in addition to fixed overheads (below the line).

Access broadcasting   Programme in which editorial decisions are made by the contributor, not by professional staff.

Acoustic   Characteristic sound of any enclosed space due to the amount of sound reflected from its wall surfaces and the way in which this amount alters at different frequencies. See also *Reverberation time, Coloration.*

Acoustic screen   Free-standing movable screen designed to create special acoustic effects or prevent unwanted sound reaching a particular microphone. One side is soft and absorbent, the other is hard and reflective.

Actuality   'Live' recording of a real event, sounds recorded on location.

Ad   Advertisement or commercial.

Ad-lib   Unscripted announcement, 'off-the-cuff' remark.

A/D converter   Analogue-to-digital converter. Creates a digital output from an analogue input, e.g. a conventional microphone recording on minidisc.

ADSL   Asymmetric Digital Subscriber Line. A digital system providing faster data transfer, and therefore greater bandwidth than ISDN.

Aerial   Device for transmitting or receiving radio waves at the point of transition from their electrical/electromagnetic form.

AGC   Automatic Gain Control. Amplifier circuit which compensates for variations in signal level, dynamic compression.

AM   Amplitude modulation. System of applying the sound signal to the transmitter frequency, associated with medium-wave broadcasting.

Analogue signal   An electrical signal that exactly represents the original shape of the acoustic or mechanical vibrations which caused it.

Anchor   Person acting as the main presenter in a programme involving several components.

AP   Associated Press. Syndicated news service.

Apple and biscuit   Microphone resembling a black ball with a circular plate fixed on one side. Omni-directional polar diagram.

ASCAP   American Society of Composers, Authors and Publishers – copyright control organisation protecting musical performance rights.

Atmosphere   Impression of environment created by use of actuality, sound effects or acoustic.

Attenuation   Expressed in decibels (dB), the extent to which a piece of equipment decreases the signal strength. Opposite of amplification.

Attenuator   Device of known attenuation deliberately inserted in a circuit to reduce the signal level.

Audience figures   Expressed as a percentage of the potential audience, or in absolute terms, the number of listeners to a single programme or sequence, daily or weekly patronage, or total usage of the station. See *Patronage, Ratings, Reach.*

Audience measurement   Research into numbers and attitudes of listeners. Methods used include: 'Aided-Recall' – person to person interview; 'Diary' – the keeping of a log of programmes heard; 'Panel' – permanent representative group reporting on programmes heard.

Audio frequency   Audible sound wave. Accepted range 20 Hz – 20 kHz.

Automatic Gain Control   See *AGC.*

Auxiliary output ('Aux')   A secondary output from a mixing desk providing a different mix independent of the main programme output, in order to send to echo, public address, foldback, etc.

Azimuth   The extent to which the gap in the recording or playback heads of a tape machine is truly vertical, i.e. at right angles to the direction of tape travel.

Back-announcement   Where the names and details of an interview or record are given immediately after the item.

Backing track   Recording of musical accompaniment heard by a soloist while adding his own performance.

Back-timing   The process of timing a live programme backwards from its intended closing time to ensure it ends on time.

Balance   Relative proportion of 'direct' to 'reflected' sound apparent in a microphone output. Also the relative volume of separate components in a total mix, e.g. voices in a discussion, musical instruments in an orchestra.

Balance control   Control for adjusting the relative volume of two stereo loudspeakers.

Bass cut   Device in microphone or other sound source which electrically removes the lower frequencies.

Bay   Standard 2145 mm × 526 mm frame housing power supplies and other technical equipment used in studios or control areas.

Bed   Instrumental backing to which words or singing are added to make a commercial or station ident.

Bias   High-frequency signal applied to the recording head of a tape machine to ensure distortion-free recording.

Bi-directional   Microphone sensitive in two directions, front and back, but completely insensitive on either side, e.g. ribbon microphone.

Black   A carbon copy, generally of typed news story.

Board   American term for studio control desk or panel.

Boom   Wheeled microphone support having a long arm to facilitate microphone placing over performers, e.g. orchestra.

Boomy   Room acoustic unduly reverberant in the lower frequencies.

Boundary effect mic   Small microphone mounted on a plate with a gap between it and the plate to give a directional polar diagram. Used on-stage for opera and theatre work. Also called Pressure Zone (PZ) effect.

Break-out box   A simple input/output unit to give easy pluggable access to a recording device, e.g. a computer.

Breakthrough   Unwanted electrical interference or acoustic sound from one source or channel affecting another.

Bulk eraser   Equipment capable of demagnetising, 'wiping' or 'cleaning' a spool of tape, or other magnetic recording, perhaps several at a time.

Burner   Colloquial term for a CD recorder, used because of its heat process.

Cans   Colloquial term for headphones.

Capacitor microphone   Microphone type based on the principle of conducting surfaces in proximity holding an electrical charge. Requires a power supply.

Capstan   The drive spindle of a tape recorder.

Cardioid   Heart-shaped area of pick-up around a microphone.

Cartridge or cart   Enclosed endless loop of tape on a single spindle which having finished is ready to start again. Also refers to Digital cart or floppy disk. Used especially for signature tunes, jingles, idents, and commercials.

Cassette   Enclosed reel to reel device of 3 mm wide tape particularly used in domestic or miniature recording machines.

CD   Compact Disc. Digital recording and playback medium.

Channel   The complete circuit from a sound source to the point in the control panel where it is mixed with others.

Chinagraph   Soft pencil used to mark tape cutting points during editing. Generally yellow.

Clean feed   A supply of cue programme in which a remote contributor hears all the programme elements other than his own. Essential to prevent howl-round in certain two-way working conditions.

Clip   A short piece of audio extracted from a longer item and illustrative of it. See also *Sound-bite*.

CODEC   A coder/decoder for converting analogue audio to a digital signal and vice versa. Used at the end of an ISDN communications line.

Coloration   Effect obtained in a room when one range of frequencies tends to predominate in its acoustic.

Compressor   Device for narrowing the dynamic range of a signal passing through it.

Condenser microphone   See *Capacitor mic.*

Control line   A circuit used to communicate engineering or production information between a studio and an outside source. Often also used as cue line. Not necessarily high quality – see *Music line.*

Copy   Written material offered for broadcast, e.g. news copy, advertising copy.

Copyright   The legal right of ownership in a creative work invested in its author, composer, publisher or designer.

Copytaster   The first reader of copy sent to a newsroom who decides whether it should be rejected or retained for possible use.

Cough key   Switch, under the speaker's control, which cuts his microphone circuit.

Crossfade   The fading in of a new source while fading out the old.

Crossplug   The temporary transposition of two circuits, normally on a Jackfield. See also *Overplugging.*

Crosstalk   Audible interference of one circuit upon another.

Cue   The pre-arranged signal to begin – visual light or gesture, verbal, musical, or scripted words.

Cue, in and **out**   The first and last words (effects or music) of a programme or item.

Cue light   A small electric light, often green, used as a cueing signal.

Cue line   A circuit used to send cue programme to a distant contributor.

Cue programme   The programme which contains a contributor's cue to start.

Cue sheet   Documentation giving technical information and introductory script for programme or insert, i.e. cue material.

Cume   Cumulating audience measurement. See *Reach.*

DAB   Digital Audio Broadcasting. Interference-free digital radio system with text, broadcast from terrestrial or satellite transmitters.

DAT   Digital Audio Tape. Sound recording and playback system in *digital mode* using small tape cassette and rotating heads – as in a videorecorder.

D/A converter   Digital-to-analogue converter. Creates an analogue output from a digital input, e.g. CD playback signal is converted in order to feed conventional loudspeakers.

dB   Decibel. Logarithmic measurement of sound intensity or electrical signal. The smallest change in level perceptible by the human ear.

Dead side of microphone   Least sensitive area.

Deferred relay   The broadcasting of a recorded programme previously heard 'live' by an audience.

Digital audio workstation   A set of digital equipment complete in itself for editing and manipulating audio material, mixing, dubbing, adding voice, making packages, etc.

Digital effects unit   Electronic equipment capable of affecting sound quality in a variety of ways, e.g. by changing frequency response, adding coloration or reverberation. Capable of creating synthetic or 'unreal' sounds.

Digital mode   The encoding of a signal as on/off pulses which represent its amplitude and frequency.

Digital signal   An electrical signal that represents its original acoustic or mechanical vibrations as a series of pulses in binary code.

Din   Plug or socket manufactured to standard of Deutsche Industrie Norm.

Directional   Property of microphone causing it to be more sensitive in one direction than in others. Also applied to transmitters, receiving aerials, loudspeakers, etc. See also *Polar diagram.*

Disc jockey   Personality presenter of record programme, generally pop music show.

Dolby system   Trade name for electronic circuitry designed to improve the signal-to-noise ratio of a programme chain.

Double-ender   Short length of audio cable with a Jack plug on each end used to connect pieces of equipment or Jacks on a Jackfield or break-out box.

Double headed   Style of presentation using two presenters.

Drive time   Periods of morning and later afternoon which coincide with commuter travel and the greatest in-car listening.

Dry run   Programme rehearsal, especially drama, not necessarily in the studio, and without music, effects or movements to mic. See *Run through.*

Drop out   Momentary drop in level or loss of quality in tape reproduction due to lack of contact between tape and repro. head.

Dub   To copy material already recorded. To make a dubbing.

Ducking unit   Automatic device providing 'voice-over' facility. See *Voice-over.*

Dynamic range   Measured in dB, the difference between the loudest and the quietest sounds.

Echo   Strictly a single or multiple repeat of an original sound. Generally refers to reverberation.

Echo plate or spring   Device for artificially adding reverberation.

Edit   The rearrangement of recorded material to form a preferred order.

Editing block   Specially shaped metal guide which holds tape in position during the cutting and splicing process.

Editorial judgement   The professional philosophy which leads to decisions on programme content and treatment.

e-mail   Electronic mail. International means of conveying computer generated correspondence or other text between similarly equipped terminals – a 'store and forward' system.

EQ   Equalisation or frequency control especially as applied to individual channels on a mixing desk.

Equity   The British Actors Equity Association. Actors' union.

Erase head   The first head of a tape recorder which cleans the tape of any existing recording by exciting it with a high frequency signal.

Fade   A decrease in sound volume. (Fade down or out.)

Fader   Volume control of a sound source used for setting its level, fading it up or down, or mixing it with other sources. Also '*Pot*'.

Fade in   An increase in sound volume. (Fade up.)

Fax   Facsimile machine, capable of sending/receiving documents, scripts and news copy via a telephone line.

Feed   A supply of programme, generally by circuit.

Feedback   See *Howl-round.*

Feedspool   Tape recorder spool which supplies the tape to the recording head. (As opposed to 'take-up' spool.)

Figure of eight   See *Bi-directional.*

Filter   Electrical circuitry for removing unwanted frequencies from a sound source, e.g. mains hum, or surface noise from an old or worn recording. Also in drama for simulating telephone or two-way radio quality, etc.

First-generation copy   A copy taken from the original recording. A copy of this copy would be a second generation copy.

Flash card   Solid-state recording medium.

Fletcher – Munson effect   The apparent decrease in the proportion of higher and lower frequencies, with respect to the middle range, as the loudspeaker listening level is decreased. Significant in correct setting of monitoring level, particularly in music balance.

Fluff   (1) Accumulation of dust on the stylus of gramophone pick-up. (2) Mistake in reading or other broadcast speech.

Flutter   Rapid variations of speed discernible in tape or disc reproduction.

FM   Frequency modulation. System of applying the sound signal to the transmitter frequency, associated with VHF broadcasting.

Foldback   Means of allowing artists in the studio to hear programme elements via a loudspeaker even while studio microphones are live.

Freelance   Self-employed broadcaster of any category – producer, contributor, operator, reporter, etc. Not on permanent full-time contract. Paid by the single contribution or over a period for a series of programmes. Non-exclusive, available to work for any employer. See also *Stringer.*

Frequency   Expressed in cycles per second or hertz, the rate at which a sound or radio wave is repeated. The note 'middle A' has a frequency of 440 Hz. A long-wave transmitter with a wavelength of 1500 metres has a frequency of 200 kHz (200 000 cycles per second). Frequency and wavelength are always associated in the formula $F \times W =$ speed. Speed is the speed of the wave, i.e. sound or radio, and in each case remains constant.

Frequency distortion   Distortion caused by inadequate frequency response.

Frequency response   The ability of a piece of equipment to treat all frequencies within a given range in the same way, e.g. an amplifier with a poor frequency response treats frequencies passing through it unequally and so its output does not faithfully reproduce its input.

Full track   Tape recording using the whole width of the tape.

Fx   Sound effects.

Gain   Expressed in decibels (dB), the amount of amplification at which an amplifier is set. Can also refer to a receiving aerial – the extent to which it can discriminate in a particular direction thereby increasing its sensitivity.

Gain control   The control which affects the gain of an amplifier, also loosely applied to any fader or volume control affecting the output level.

Gramophone, gramdeck or **grams**   Turntable and associated equipment for the reproduction of records.

GTS   Greenwich Time Signal – six pips ending at a precise time.

Gun-mic   Microphone resembling a long-barrelled shotgun. Highly directional, used for nature recordings or where intelligibility is required at some distance from the sound source, e.g. OBs.

Half track   Tape recording erase and recording applied only to the 'top' half of the tape, as opposed to 'full track' which uses the whole width.

Hammocking   Scheduling term referring to the need to support a low audience or specialist programme by placing more popular material before and after it in order to maintain a strong average listening figure.

Handout   Press information or publicity sheet issued to draw attention to an event.

Hand signals   System of visual communication used through the glass window between a studio and its control area, or in a studio with a 'live' mic. See *Wind-up.*

Harmonic distortion   The generation of spurious upper frequencies.

Head amplifier   Small amplifier within a microphone, especially capacitor type.

Head gap   Narrow vertical slot at the front of tape recorder erase, record and replay heads.

Headline   Initial one-sentence summary of news event.

Hertz   Hz. Unit of frequency, one complete cycle per second.

Hiss   Unwanted background noise in the frequency range 5 – 10 kHz, e.g. tape hiss.

Howl-round   Acoustic or electrical positive feedback generally apparent as a continuous sound of a single frequency. Often associated with public address systems. Avoided by decreasing the gain in the amplifying circuit, cutting the loudspeaker or in contribution working through the use of a clean feed circuit.

Hum   Low frequency electrical interference derived from mains power supply.

Hypercardioid   A cardioid microphone having a particularly narrow angle of acceptance at its front which decreases rapidly towards the sides.

ID   Station identification or ident.

IFPI   International Federation of Phonographic Industries. International organisation of record manufacturers to control performance and usage rights.

Igranic jack   Jack plug providing two connections.

In cue   The first words of a programme insert, known in advance. Can also be music.

Insert   A short item used in a programme, e.g. a 'live' insert, a tape insert.

Intercom   Local voice communication system.

Internet   Worldwide digital communication system linking similarly equipped computers. See also *e-mail, Website.*

IPS   Inches per second. Tape-recording term, refers to speed of tape travel past the recording and replay heads.

ISDN   Integrated Services Digital Network. A system of conveying high quality digital audio signals over the public telephone system. See also *CODEC.*

Jack   Socket connected to an audio circuit. Can incorporate a switch activated by insertion of Jack plugs – a 'break' Jack.

Jackfield or **patch panel**   Rows of Jacks connected to audio sources or destinations. Provides availability of all circuits for interconnection or testing.

Jack plug or **post office Jack**   Plug type used for insertion in Jack socket comprising three connections, a circuit pair plus earth, known as 'ring tip and sleeve'. See also *Double ender.*

Jingle   Short musical item used as station ID, or in advertisement.

Jingle package   The set of jingles used by a station to establish its audio logo.

Jock   See *Disc jockey.*

Jog   A control on CD and DAT recorders enabling the accurate location of any point in the recording.

Joining tape   Adhesive tape used in tape editing.

Key   Switch.

Kilo   Thousand. Kilohertz – frequency in thousands of cycles per second. Kilowatt – electrical power, a thousand watts.

Landline   See *Line.*

Lavalier microphone   Small microphone hung round the neck **(lanyard mic)** or fastened to clothing.

Lazy arm   Small boom-type microphone stand suitable for suspending a microphone over a 'talks' table.

Lead sheet   Basic musical score indicating instrumentation of melody. Used for microphone control during music balance.

Lead story   The first, most important story in a news bulletin.

Leader   Inert coloured tape having the same dimensions as recording tape spliced into a recorded programme or insert to give visual indication of beginnings and endings. Before the beginning – white or yellow; intermediate spacers – yellow; after the end – red.

Level   (1) A test prior to recording or broadcasting to check the volume of the speaker's voice – 'take some level'. (2) Expressed in dBs, plus or minus, the measurement of electrical intensity against an absolute standard, zero level (1 MW in 600 ohms).

Limiter   Device to prevent the signal level exceeding a pre-set value.

Line   Physical circuit between two points for programme or communication purposes.

Line equalisation   The process which compensates for frequency distortion at the receiving end of a landline.

Line-up   Technical setting up of circuits to conform to engineering standards. Line-up tone of standard frequency and level used to check the gain of all component parts.

Lip mic   Noise-excluding ribbon microphone designed for close working, e.g. OB commentary.

Log   Written record of station output. Can also be recorded audio.

'M' signal   The combination of left and right stereo signals, i.e. the mono signal.

Marching box   Sound effects device comprising small box partially filled with gravel used to simulate marching feet.

MCPS   Mechanical Copyright Protection Society. Organisation which controls the copying or dubbing of copyright material.

Mega   Million. Megahertz – frequency in millions of cycles per second. Megawatt – electrical power, a million watts.

Microphonic   Faulty piece of electronic equipment sensitive to mechanical vibration – acting like a microphone.

Middle of the road   Popular, mainstream music with general appeal. Non-extreme.

Minidisc (MD)   Digital recording and playback system using a 64 mm disc in portable recorders and studio decks.

Mixer   Studio, OB, or PA control desk for mixing together sound sources to the appropriate level.

Modem   Modulator/demodulator. Converts an (acoustic) analogue signal to digital signal and vice versa. Used to send computer output to a telephone line.

Modulation   Variations in a transmission or recording medium caused by the presence of programme. Often abbreviated to Mod.

Module   Interchangeable equipment component.

Montage   Superimposition of sounds and/or voices to create a composite impression.

MOR   Middle of the road music.

MU   Musicians Union.

Multi-tracking   Two or more audio tracks are recorded separately and subsequently mixed for the final result.

Music line   High-quality landline or satellite circuit suitable for all types of programme, not only music. Compare with *Control line.*

NAB   National Association of Broadcasters. American trade organisation which secures agreement on standards of procedure and equipment, e.g. NAB spool, a professional tape reel type.

Nagra (Trade name)   High-quality portable tape recorder.

Noise   Extraneous sound, electrical interference, or background to a signal.

Noise gate   Device which allows a signal to pass through it only when the input level exceeds a pre-set value.

OB   Outside broadcast.

Off-mic   A speaker or other sound source working outside a microphone's most sensitive area of pick-up. Distant effect due to drop in level and greater proportion of reflected to direct sound.

Omni-directional   A microphone sensitive in all directions. Also applied to transmitters and aerials.

One-legged   'Thin' low-level quality resulting from a connection through only one wire of a circuit pair.

Open-ended   A programme without a pre-determined finishing time.

Optimod   Audio compressor to maximise the modulation of a transmitter in order to obtain optimum signal strength.

Out cue   Final words of a contribution, known in advance, taken as a signal to initiate the following item in a sequence.

Out of phase   The decrease in level and effect on quality when two similar signals are combined in such a way as to cancel each other.

Outside source   Programme originating point remote from the studio, or the circuit connection from it.

Overload distortion   The distortion suffered by a programme signal when its electrical level is higher than the equipment can handle. When this happens non-continuously it is referred to as 'peak distortion'. Also referred to as 'squaring off'.

Overplugging   The substitution of one circuit for another by the insertion of Jack plug in a break Jack.

PA   (1) Press Association. News Agency. (2) Public address system.

Par   Paragraph. Journalist's term often applied to news copy.

P as B   Programme as broadcast. Documentation giving complete details of a programme in its final form – duration, inserts, copyright details, contributors, etc.

Package   Edited programme or insert offered complete with links ready for transmission.

Panel   Studio mixing desk, control board or console.

Pan-pot Panoramic potentiometer. Control on studio mixing desk which places a source to the left or right in a stereo image.

Parabolic reflector Microphone attachment which focuses sound waves thereby increasing directional sensitivity. Used for OBs, nature recordings, etc.

Parallel strip An inert row of jacks mounted on a Jackfield not connected to any other equipment but connected in parallel to each other. Used for joining programme sources together, connecting equipment, or multiplying outputs from a single feed. Also available on its own in the form of a 'Junction Box'.

Patch panel See *Jackfield*.

Patronage See *Reach*.

Peak distortion See *Overload distortion*.

Peak programme meter Voltmeter with a slugged slow decay time, designed to indicate levels and peaks of electrical intensity for the purposes of programme control.

Phantom power Method of providing a working voltage to a piece of equipment, e.g. a microphone, using the programme circuit and earth (ground).

Phase distortion The effect on the sound quality caused by the imprecise combination of two similar signals not exactly in phase with each other.

Pick-up Gramophone record reproducing components which convert the mechanical variations into electrical energy, pu-arm, pu-head, pu-shell, pu-stylus.

Pilot Programme to test the feasibility of, or gain acceptance for, a new series or idea.

Pinch roller Rubber wheel which holds tape against tape recorder drive capstan.

Plug Free advertisement.

Polar diagram Graph showing the area of a microphone's greatest sensitivity. Also applies to aerials, transmitters and loudspeakers. Directivity pattern.

Popping Descriptive term applied to 'mic-blasting', the effect of vocal breathiness close to microphone.

Post-echo The immediate repeating at low level of sounds replayed from a tape recording. See *Print-through*.

Post office jack See *Jack plug*.

Pot Potentiometer. See *Fader*.

Pot cut The cutting off of a recording during replay before it has finished by closing its fader – generally to save time. 'Instant editing'.

PPL Phonographic Performance Ltd. Organisation of British record manufacturers to control performance and usage rights.

PPM See *Peak programme meter*.

Prefade The facility for hearing and measuring a source before opening its fader, generally on a studio mixing desk.

Prefade to time The technique of beginning an item of known duration before it is required so that it finishes at a precise time, e.g. closing signature tune. Also known as back-timing.

Presence A sense of 'realistic closeness' often on a singer's voice. Can be aided by boosting the frequencies in the range 2.8 kHz to 5.6 kHz.

Prime time The best, most commercial hours of station output, e.g. 6.30 a.m. to 10.30 a.m.

Print-through The reproduction at low level of recorded programme through the magnetic interaction of layers of tape due to their close proximity while wound on a spool. The cause of post- and pre-echo. Often the result of tight winding through spooling at high speed, and storage of recorded tape at too high a temperature.

Producer The person in charge of a programme and responsible for it.

Promo On-air promotion of station or programme. Also *Trail*.

PRS Performing Right Society. Organisation of authors, composers and publishers for copyright protection.

PSA Public Service Announcement – made by the station in the public interest, or on behalf of a charity or other non-commercial body – or at its own discretion for a private individual – for which no charge is made.

Puff See *Plug*.

PZ (Pressure Zone) mic See *Boundary effect mic*.

Q & A Question and Answer basis of a discussion between a programme presenter and a specialist correspondent. Less formal than an interview.

Quad Quadraphony or Quadrasonic. Four channel sound reproduction providing front and rear, left and right coverage.

Radio mic Microphone containing or closely associated with its own portable transmitter. Requires no cable connection, useful for stage work, OBs, etc.

Ratings   Audience measurement relating to the number of listeners to a specific programme.

RDS   Radio Data System. A data signal added to FM and digital transmissions, used for carrying text, e.g. station ident and other messages, for display by the receiver. Also provides automatic switching of car radios for local traffic information, retuning for best signal, encoding of programmes for recording, etc.

Reach   Term used in audience measurement describing the total number of different listeners to a station or service within a specified period. Most often expressed as a percentage of the potential audience. Weekly Reach. Also *Market Penetration, Patronage.*

Record head   The part of a tape machine which converts the electrical signal into magnetic variations and transfers them to the tape.

Reduction   Playback of a multi-track music recording to arrive at a final mix. Also mix-down.

Rehearse-record   Procedure most used in music recording or drama for perfecting and recording one section before moving on to the next.

Relay   (1) Simultaneous transmission of a programme originating from another station. (2) Transmission of a programme performed 'live' in front of an audience. See also *Deferred relay.* (3) Electrically operated switch.

Repro head   The part of a tape machine which converts the magnetic pattern on the tape into electrical signal. Reproduction or playback device.

Residual   Artist's repeat fee.

Reverberation   The continuation of a sound after its source has stopped due to reflection of the sound waves.

Reverberation time   Expressed in seconds, the time taken for a sound to die away to one millionth of its original intensity.

Reverse talkback   Communications system from studio to control cubicle.

Ribbon microphone   A high-quality microphone using electromagnetic principle. Bi-directional polar diagram.

Rip 'n' read   News bulletin copy sent from a central newsroom designed to be read on the air without rewriting.

ROT   Recording Off Transmission. Recording made at the time of transmission, not necessarily off-air.

RSA   Response Selection Amplifier – Device for control of treble and bass frequencies, and 'presence'. See *EQ.*

Running order   List of programme items and timings in their chronological sequence.

Run through   Programme rehearsal.

'S' signal   The difference between left and right stereo signals, i.e. the stereo component.

Satellite studio   Small outlying studio, perhaps without permanent staff but capable of being used as a contribution point via a link with the main studio/station centre.

SB   Simultaneous Broadcast. Relay of programme originating elswhere. Conveyed from point to point by system of permanent SB lines, or taken 'off-air'.

Script   Complete text of a programme or insert from which the broadcast is made.

Segue   The following of one item immediately on another without an intervening pause or link. Especially two pieces of music.

Share   Audience measurement term describing the amount of listening to a specified station or service expressed as a percentage of the total listening to all services heard in that area.

Sibilance   An emphasis on the 's' sounds in speech. May be accentuated or reduced by type and position of microphone.

Sig tune   Signature tune. Identifying music at the beginning and end of a programme or regular insert.

Slug   Short identifying title given to a short item, particularly a news insert. Also catchline.

Solid state   Transistorised or integrated circuitry as opposed to that containing valves.

Sound-bite   A short piece of audio said to sum up a particular truth or point of view, able to stand alone.

Spot Fx   Practical sound effects created live in the studio.

Spin-doctor   Organisation person hired to promote positive aspects of policy and events, and to suppress the negative.

Squelch   Means of suppressing unwanted noise in the reception of a radio signal. See *Noise gate.*

Sting   Single music chord, used for dramatic effect.

Stock music   In-house library of recorded music.

Stringer   Freelance contributor paid by the item. Generally a journalist at outlying place not covered by staff.

Stylus   Small diamond tipped arm protruding from gramophone pick-up. In contact with the record surface it conveys the mechanical vibrations to the cartridge for conversion into electrical energy.

Sustaining programme   Programme supplied by a syndicating source or elsewhere to maintain an output for a station making its own programmes for only part of the day.

Sweep   The process of audience survey for a particular station or service within a given timescale.

Sync output   Programme replayed from the record heads of a multi-track recording machine heard by performers while they record further tracks.

Talkback   Voice communication system from control cubicle to studio or other contributing point.

Talks table   Specially designed table for studio use, often circular with an acoustically transparent surface and a hole in the middle to take a microphone.

Tape   Magnetic recording material.

Tapes   Journalist's term for copy received by teleprinter or paper strips on which such material is printed.

TBU   Telephone Balance Unit. Interface device used in conjunction with calls on a phone-in programme. Minimises risk of howl-round while providing pre-set level of caller to studio presenter and vice versa. Isolates public telephone equipment from broadcaster's equipment.

Telex   Teleprinter system of written communication.

Tie line   Any circuit pair connecting two programme areas, especially within the same building.

Tone   A test or reference signal of standard frequency and level. For example, 1 kHz at 0 dB.

Top and tail   A shortened programme rehearsal where only the openings and closing of inserts are played. Also, adding the opening and closing to a package.

Tracking weight   The downward pressure of a gramophone pick-up transmitted through its stylus.

Trail   Broadcast item advertising forthcoming programme. On-air promotion or 'promo'.

Transcript   The text of a broadcast as transmitted, often produced from an off-air recording.

Transcription   A high-quality recording of a programme, often intended for reproduction by another broadcasting service.

Transducer   Any device which converts one form of energy into another, e.g mechanical to electrical, acoustic to electrical, electrical to magnetic, etc.

Transient response   The ability of a microphone or other equipment to respond rapidly to change of input or brief energy states.

Two-way   Discussion or interview between two studios remote from each other. Also an interview of a specialist correspondent by a programme presenter. See also *Q & A*.

Uher   Trade name of portable tape recorder.

UHF   Ultra High Frequency. Radio or television transmission in the range of frequencies from 300 MHz to 3000 MHz.

UPI   United Press International. Syndicated news service.

VHF   Very High Frequency. Radio or television transmission in the range of frequencies from 30 MHz to 300 MHz.

Voice-over   Voiced announcement superimposed on lower level material, generally music.

Voice report   Broadcast newspiece in the reporter's own voice.

Vox pop   'The voice of the people'. Composite recording of 'street' interviews.

VU meter   Volume Unit meter calibrated in decibels measuring signal level especially as a recording level indicator.

Warm-up   Initial introduction and chat, generally by a programme presenter or producer, designed to make an audience feel at home and create the appropriate atmosphere prior to a live broadcast or recording.

Wavelength   Expressed in metres, the distance between two precisely similar points in adjacent cycles in a sound or radio wave. The length of one cycle. Used as the tuning characteristic or 'radio address' of a station. See also *Frequency*.

Website   An organisation's or individual's location on the Internet for promotional publicity, information, sales, etc. Not generally the address for correspondence.

Wild or **wild track**   Term borrowed from film to describe the recording of atmosphere, actuality or effects at random without a precise decision on how they are to be used in a programme.

Windshield   Protective cover of foam rubber, plastic, or metal gauze, designed to eliminate wind noise from microphone. Essential on outdoor use or close vocal work.

Wind-up   Signal given to broadcaster to come to the end of his programme contribution. Often by means of index finger describing slow vertical circles, or by flashing cue light.

Wipe   To erase tape.

Wow   Slow speed variations discernible in tape or disc reproduction.

Wrap   A short piece of actuality audio 'wrapped around' by a vocal introduction and a back-announcement. Frequently used in news bulletins.

Zero level tone   A standard reference level signal (0 dB or 1 mW in 600 ohms at a frequency of 1 kHz) used to line up broadcasting equipment.

Further reading – a selection

Historical survey and reference

Allen, J., *Careers in Television and Radio*. Kogan Page, 1988

Barnard, S., *On the Radio; Music Radio in Britain*. Open University Press, 1989

Barnouw, E., *A History of Broadcasting in the United States, vols 1–3. A tower in Babel, to 1933; The golden web, 1933–53; The image empire, 1950 onward*. New York: OUP, 1966–70

Baron, M., *Independent Radio: the story of independent radio in the United Kingdom* (new edn). Dalton, 1976

Blumler, J. G., *Broadcasting Finance in Transition: a comparative handbook*. OUP, 1991

Boemer, M. L., *The Children's Hour: radio programs for children 1929–1956*. Scarecrow, 1989 (USA)

Briggs, A., *The History of Broadcasting in the United Kingdom, vols 1–4. The birth of broadcasting; The golden age of wireless; The war of words; Sound and vision*. OUP, 1961–79

Briggs, A., *The BBC: the first fifty years*. OUP, 1985

Browne, D. R., *International Radio Broadcasting*. Praeger, 1982

Burns, T., *The BBC – public institution and private world*. Macmillan, 1977

Butler, B. (ed.), *Sports Report: 40 years of the best*. Macdonald, 1987

Cain, J., *The BBC: 70 years of broadcasting*. BBC, 1992

Carpenter, H., *The Envy of the World: 50 years of the BBC Third Programme and Radio 3*. Phoenix-Orion, 1997

Chapman, R., *Selling the Sixties: The pirates & pop music radio*. Routledge, 1992

Commonwealth Broadcasting Directory. Commonwealth Broadcasting Association

Douglas, G., *The Early Days of Radio*. McFarland, 1987 (USA)

Gifford, D., *The Golden Age of Radio: an illustrated companion*. Batsford, 1985

Head, S., *Broadcasting in America: a survey of television and radio*, 4th edn. Boston: Houghton Mifflin, 1982

Hind, J. and Mosco, S., *Rebel Radio*. Pluto, 1985

Keith, M., *Selling Radio Direct*. Focal Press, 1992

Kenyon, N., *The BBC Symphony Orchestra: the first fifty years 1930–1980*. BBC, 1981

Leapman, M., *The Last Days of the Beeb*. Allen & Unwin, 1986

List, D., *Audience Survey Cookbook*. Adelaide, Australian Broadcasting Corp., 1997

McCavitt, W., *Broadcasting Around the World*. TAB Books Inc., 1981

MacDonald, B., *Broadcasting in the United Kingdom – a guide to information sources*. Mansell, 1988

McIntyre, I., *The Expense of Glory: A life of John Reith*. HarperCollins, 1993

Mansell, G., *Broadcasting to the World: forty years of the external services*. BBC, 1973

Mansell, G., *Let Truth be Told: 50 years of BBC external broadcasting*. Weidenfeld, 1982

Martin-Jenkins, C., *Ball by Ball: the story of cricket broadcasting*. Grafton, 1990

Mickelson, S., *America's Other Voice: Radio Free Europe and Radio Liberty*. New York: Praeger, 1983

Milne, A., *The Memoirs of a British Broadcaster*. Hodder & Stoughton, 1988

Milner, R., *Reith and the BBC years*. Mainstream, 1983

Mytton, G. (ed.), *Global Audiences: research for Worldwide Broadcasting*. John Libbey, 1993

Mytton, G., *Handbook on Radio & Television Audience Research*. BBC, UNICEF and UNESCO, 1999

Partner, P., *Arab Voices: the BBC Arabic Service 1938–1988*. BBC, 1988
Paulu, B., *Television and Radio in the United Kingdom*. Macmillan, 1981
Pegg, M., *Broadcasting and Society, 1918–39*. Croom Helm, 1983
Reeves, G., *Communications and the Third World*. Routledge, 1993
Reid, C., *Action Stations: a history of Broadcasting House*. Robson Books, 1987
Robinson, J., *Learning over the Air: sixty years of partnership in adult learning*. BBC, 1982
Russell, F., *Getting Jobs in Broadcasting*. Cassell, 1989
Rusling, P., *The Lid off Laser 558*. Pirate, 1984
Took, B., *Laughter in the Air: an informal history of radio comedy* (revised edn). Robson Books, 1982
Tusa, J., *A World in Your Ear*. Broadside Books, 1992
Wedell, G. (ed.), *Making Broadcasting Useful – the African experience*. Manchester University Press, 1986
Whitehead, K., *The Third Programme: a literary history*. Clarendon, 1989
Wolfe, K., *The Churches and the BBC 1922–1956: the politics of broadcast religion*. SCM Press, 1984
Wood, J., *History of International Broadcasting*. Peter Peregrinus, 1992

Broadcasting and society

Adams, V., *The Media and the Falklands Campaign*. Macmillan, 1986
Barendt, E., Lustgarten, L., Nome, K. and Stephenson, H., *Libel and the Media*. Clarendon Press, 1997
Barnett, S., *The Listener Speaks: the radio audience and the future of radio*. HMSO, 1989
Boyd-Barrett, O. and Braham, P. (eds), *Media, Knowledge and Power*. Croom Helm, 1987
Crone, T., *Law and the Media* (3rd edn). Focal Press, 1995
Curran, J., Smith, A. and Wingate, P. (eds), *Impacts and Influences: essays on media power in the 20th century*. Methuen, 1987
Donahue, H., *The Battle to Control Broadcast News*. MIT Press, 1989
Gans, H., *Deciding What's News*. New York: Pantheon, Random House, 1979. London: Vintage, 1980 (op); Constable, 1980
Hannan, P., *Wales on the Wireless: a broadcasting anthology*. Gomer/SSC Wales, 1988
Hawthorn, J., *Reporting Violence: lessons from Northern Ireland?* BBC, 1981
Hoggart, R. and Morgan, J. (eds), *The Future of Broadcasting: essays on authority, style and choice*. Macmillan, 1982
Hosley, D., *Hard News: women in broadcast journalism*. Greenwood, 1987
Hybels, S. and Ulloth, D., *Broadcasting: introduction to radio and television*. Van Nostrand Reinhold, 1978
Koch, T., *The News as Myth: fact & context in journalism*. Greenwood, 1990
Kuhn, R. (ed.), *The Politics of Broadcasting*. Croom Helm, 1985
Legat, M., *The Writer's Rights*. A & C Black, 1995
Lewis, P. M., *The Invisible Medium: public, commercial, and community radio*. Macmillan, 1989
Luther, S., *The United States and the Direct Broadcast Satellite: the politics of international broadcasting in space*. OUP, 1988
MacCabe, C. and Stewart, O. (eds), *The BBC and Public Service Broadcasting*. Manchester University Press, 1986
MacDowall, I., *Reuters Handbook for Journalists*. Butterworth-Heinemann, 1992
Page-Herweg, G and A., *Radio's Niche Marketing Revolution*. Focal Press, 1997
Patterson, M., *Behind the Lines: case studies in investigative reporting*. Columbia University Press, 1986
Rolston, W., *The Media and Northern Ireland*. Macmillan, 1991
Spark, S., *A Journalist's Guide to Sources*. Focal Press, 1996
Schlesinger, P., *Putting Reality Together: BBC news*. Methuen, 1987
Swann, M., *Freedom and Restraint in Broadcasting: the British experience*. BBC, 1975
Taylor, L., *Uninvited Guests*. Chatto & Windus, 1986
Wedell, G. and Katz, E., *Third World*. Macmillan, 1978
West, W. J., *Truth Betrayed (European radio services)*. Duckworth, 1987
Willis, J., *The Shadow World: life between the news media and reality*. Praeger, 1991

Programmes and personalities

Adams, D., *The Hitchhiker's Guide to the Galaxy.* Pan Books, 1992
Baehr, H. and Ryan, M., *Shut Up and Listen: women and local radio – a view from the inside.* Comedia, 1984
Bennett, A., *Talking Heads – Six Classic Monologues.* BBC, 1998
Day, Sir R., *Grand Inquisitor: memoirs.* Weidenfeld & Nicolson, 1989; pbk Pan Books, 1990
Day, Sir R., *But With Respect.* Weidenfeld & Nicolson, 1993
Deayton, A., *Radio Active Times (based on the BBC Radio 4 series).* Sphere, 1986
Drakakis, J., *British Radio Drama.* CUP, 1981
Flywheel, *Shyster and Flywheel: Marx Brothers' lost radio show.* Chatto & Windus, 1989; Pantheon, NY, 1988
Francey, D., *And its All Over Sports Programming.* John Donald, 1988
Gallagher, J., *The Archers Omnibus.* BBC Books, 1990
Galton, R., *Hancock's Half Hour Radio Scripts.* BBC, 1987
Garfield, S., *The Nation's Favourite: The True Adventures of Radio 1.* Faber & Faber, 1998
Hall, L., *Spoonface Steinberg & Other Plays from Radio 4.* BBC, 1997
Halper, D., *Full-service Radio: programming for the community.* Focal Press, 1991
Hamilton, D., *The Music Game.* W. H. Allen, 1986
Hancock, F. and Nathan, D., *Hancock.* BBC, 1996
Hitchcock, J., *Sportscasting.* Focal Press, 1991
Johnston, B., *It's a Funny Game.* Crescent, 1986
Johnston, B., *View from the Boundary.* BBC, 1990
Jones, D., *Microphones & Muddy Boots: a journey into natural history broadcasting.* David & Charles, 1987
Lawley, S., *Sue Lawley's Desert Island Discussions.* Hodder & Stoughton, 1990
Lewis, P. (ed.), *Radio Drama.* Longman, 1981
Michelmore, C. and Metcalf, J., *Two-way Story.* Hamilton, 1986; Futura, 1987
Milligan, S., *The Lost Goon Shows.* Robson, 1987
Moore, R., *Tomorrow is Too Late.* Constable, 1988
Nicholson, M., *A Measure of Danger: memoirs of a British war correspondent.* HarperCollins, 1991
Popham, M. (ed.), *From Our Own Correspondent.* Broadside Books, 1991
Priestland, G., *Yours Faithfully.* Collins, 1979; pbk 1981
Priestland, G., *The Unquiet Suitcase: Priestland at sixty.* Deutsch, 1988
Priestland, G., *Something Understood.* Deutsch, 1986
Reid, C., *Action Stations.* Robson, 1987
Roberts, P. (ed.), *Plays without Wires.* Sheffield Academic Press, 1989
Rodger, I., *Radio Drama.* Macmillan, 1981
Simpson, J., *The Darkness Crumbles: Despatches from the Barricades.* Hutchinson, 1992
Smethurst, W., *The Archers: the new official companion.* Weidenfeld and Nicolson, 1987
Smith, G., *Wogan.* W. H. Allen, 1987
Sperber, A. M., *Murrow – His Life and Times.* Michael Joseph, 1987
Spink, G. (ed.), *The Best of Our Own Correspondent.* I. B. Tauris, 1993
Talbot, G., *Ten Seconds from Now: a broadcaster's story.* Hutchinson, 1973
Timpson, J., *The Lighter Side of 'Today'.* Allen and Unwin, 1983; pbk 1985
Took, B. (ed.), *The Best of 'Round the Horne'.* Equation, 1989
Toye, J., *The Archers: Family Ties 1951–1967: The Early Years.* BBC, 1998
Wertheim, A., *Radio Comedy.* OUP, 1979
Whitburn, V., *The Archers: The Official Inside Story.* Virgin, 1992
Best Radio Plays of 1986. Methuen/BBC, 1987.
Young Playwrights Festival 1988: BBC radio drama. BBC, 1988

Radio production

Ahamed, U. and Grimmett, G. (eds), *Educational Broadcasting – Radio.* Asia – Pacific Institute for Broadcasting Development, 1979
Ash, W., *The Way to Write Radio Drama.* Elm Tree Books, 1985
Aspinall, R., *Radio Programme Production: a manual for training.* Paris: UNESCO, 1973
A Voice for Everyone: all you need to know about community radio. Dublin Veritas, 1988
BBC Producers Guidelines. BBC, 1996

Bland, M., *You're on Next! How to survive on television and radio.* Kogan Page, 1985
Bliss, E. and Patterson, J., *Writing News for Broadcast.* Columbia UP, 1978
Boyd, A., *Broadcast Journalism: techniques of radio and TV news (4th edn).* Focal Press, 1997
Brand, J., *Hello, Good Evening and Welcome: a guide to being interviewed on television and radio* (2nd edn). Shaw, 1985
Burchfield, R. W., *The Spoken Word: a BBC guide.* BBC, 1981
Chantler, P. and Harris, S., *Local Radio Journalism.* Focal Press, 1996
Cohler, D. K., *Broadcast Journalism: a guide for the presentation of radio & television news.* Prentice-Hall, 1986
Coleman, H. W., *Case Studies in Broadcast Management.* New York: Hastings House, 1978
Crissell, A., *Understanding Radio.* Routledge, 1994
Ditingo, V., *The Remaking of Radio.* Focal Press, 1995
Gage, L., *Guide to independent radio journalism.* Duckworth, 1990
Gage, L., Douglas, L. and Kinsey, M., *Commercial Radio Journalism.* Focal Press, 1999
Gooch, S., *Writing a Play* (2nd edn). A & C Black, 1995
Greenwood, W. and Welsh, T. (eds), *McNae's Essential Law for Journalists* (9th edn). Butterworths, 1985
Hall, M., *Broadcast Journalism: an introduction to news writing.* New York: Hastings House, 1978
Hall, R., *Writing Your First Play.* Focal Press, 1998
Halper, D., *Radio Music Directing.* Focal Press, 1991
Hartley, J., *Understanding News.* Methuen, 1982
Hasling, J., *Fundamentals of Radio Broadcasting.* McGraw-Hill, 1980
Hawkridge, D. and Robinson, J., *Organising Educational Broadcasting.* Croom Helm, 1981
Hilliard, R. (ed.), *Radio Broadcasting: introduction to the sound medium* (3rd edn). Longman, 1985
Horstmann, R., *Writing for Radio* (3rd edn). A & C Black, 1997
Hudson, R., *Inside Outside Broadcasts.* R. & W. Publications Ltd, 1993
Kay, M. and Popperwell, A., *Making Radio.* Broadside Books, 1992
Keith, M., *Radio Programming: consultancy and formatics.* Focal Press, 1987
Keith, M., *Broadcast Voice Performance.* Focal Press, 1989
Keith, M., *Radio Production: art and science.* Focal Press, 1990
Keith, M., *The Radio Station* (4th edn). Focal Press, 1997
MacLoughlin, S., *Writing for Radio.* How to Books, 1998
Millar, C. J., *Contempt of Court* (2nd edn). Clarendon Press, 1989
Miller, G. M. (revised by Pointon, G.), *BBC Pronouncing Dictionary of British Names* (2nd edn). OUP, 1983
Mott, R., *Sound Effects: Radio, TV, and Film.* Focal Press, 1990
Quaal, W. and Brown, J., *Broadcast Management* (2nd edn). New York: Hastings House, 1976
Redfern, B., *Local Radio.* Focal Press, 1979
Reese, D. and Gross, L., *Radio Production Worktext* (3rd edn). Focal Press, 1998
Routt, E. *et. al.,* *The Radio Format Conundrum.* New York: Hastings House, 1978
Sheppard, R., *The DJ's Handbook: from scratch to stardom.* Javelin Books, 1986
BBC English Dictionary. HarperCollins, 1992
BBC Pronunciation: policy and practice. BBC Pronunciation Unit, 1974
Writing for the BBC: a guide for writers on possible markets within the BBC (7th edn). BBC, 1983

Technical matters

Aldred, J., *Manual of Sound Recording* (3rd edn). Dickson Price, 1982
Alkin, G., *Sound Recording and Reproduction* (3rd edn). Focal Press, 1996
Bartlett, B., *Stereo Microphone Techniques.* Focal Press, 1991
Borwick, J., *Sound Recording Practice* (3rd edn). OUP, 1992
Borwick, J., *Microphones: Technology & Technique.* Focal Press, 1990
Diamant, L., *The Broadcast Communications Dictionary* (2nd edn). Dickson Price, 1982
Hellyer, H. W. and Sinclair, I. R., *Questions and Answers on Radio and Television* (4th edn). Newnes, 1979
King, G. J., *Beginner's Guide to Radio* (9th edn). Heinemann, 1984
McCavitt, W. E. and Pringle, P. K., *Electronic Media Management.* Focal Press, 1986
Nallawalla, A., Cushen, A. and Clark, B., *Better Radio/TV Reception.* Ashley Publishing, Australia, 1986
Nisbett, A., *The Use of Microphones* (3rd edn). Focal Press, 1989

Nisbett, A., *The Sound Studio* (6th edn). Focal Press, 1995
Oringel, R. S., *Audio Control Handbook* (5th edn). New York: Hastings House, 1983
Roberts, R. S., *Dictionary of Audio, Radio and Video*. Butterworths, 1981
Ross, J. F., *Handbook for Radio Engineering Managers*. Butterworths, 1980
Runstein, R. and Huber, D. M., *Modern Recording Techniques*. Howard Sams & Co., 1986
Talbot-Smith, M., *Broadcast Sound Technology* (2nd edn). Focal Press, 1995
Talbot-Smith, M., *Sound Assistance*. Focal Press, 1997
Watkinson, J., *An Introduction to Digital Audio*. Focal Press, 1994
Withers, D., *Radio spectrum management*. Peregrinus for the IEE, 1991

Other bibliography sources

British Broadcasting 1922–1982 – a select bibliography. BBC, 1983
Catalogue of radio and television training materials from the United Kingdom. British Council, 1977

Index

www.focalpress.com

Visit our web site for:

- the latest information on new and forthcoming Focal Press titles
- special offers
- our e-mail news service

Join our Focal Press Bookbuyers' Club

As a member, you will enjoy the following benefits:

- special discounts on new and best-selling titles
- advance information on forthcoming Focal Press books
- a quarterly newsletter highlighting special offers
- a 30-day guarantee on purchased titles

Membership is free. To join, supply your name, company, address, telephone/fax numbers and e-mail address to:
Elaine Hill
E-mail: elaine.hill@repp.co.uk
Fax: +44 (0) 1865 314572
Address: Focal Press, Linacre House, Jordan Hill, Oxford OX2 8DP

Catalogue

For information on all Focal Press titles, we will be happy to send you a free copy of the Focal Press Catalogue.

Tel: 01865 314220
E-mail: Jo.Blackford@repp.co.uk

Potential authors

If you have an idea for a book, please get in touch:

Europe and rest of world
E-mail: focal.press@repp.co.uk

USA
E-mail: editors@focalpress.com